Plant Layout and Material Handling

Fred E. Meyers
Associate Professor
Department of Technology
Southern Illinois University at Carbondale

REGENTS/PRENTICE HALL
Englewood Cliffs, NJ 07632

Library of Congress Cataloging-in-Publication Data

Meyers, Fred E.
    Plant layout and material handling / Fred E. Meyers.
        p.   cm.
    Includes bibliographical references and index.
    ISBN 0-13-013475-9
    1. Plant layout.   2. Materials handling.   I. Title.
TS178.M48   1993
658.5′1—dc20   92–41485
                                                    CIP

*To the industrial technology students
of Southern Illinois University—Carbondale*

Acquisition Editor: Rob Koehler
Production Editors: Fred Dahl and Rose Kernan
Copy Editor: Rose Kernan
Cover Designer: Marianne Frasco
Designers: Fred Dahl and Rose Kernan
Prepress Buyer: Ilene Levy
Manufacturing Buyer: Ed O'Dougherty
Supplements Editor: Judy Casillo

 © 1993 by REGENTS/PRENTICE HALL
A Division of Simon & Schuster
Englewood Cliffs, New Jersey 07632

Printed in the United States of America
10 9 8 7 6 5 4 3 2 1

ISBN 0-13-013475-9

Prentice-Hall International (UK) Limited, London
Prentice-Hall of Australia Pty. Limited, Sydney
Prentice-Hall Canada Inc., Toronto
Prentice-Hall Hispanoamericana, S.A., Mexico
Prentice-Hall of India Private Limited, New Delhi
Prentice-Hall of Japan, Inc., Tokyo
Simon & Schuster Asia Pte. Ltd., Singapore
Editora Prentice-Hall do Brasil, Ltda., Rio de Janeiro

Contents

The Material Handling Problem Solving Procedure, 170
Material Handling Checklist, 171
Questions, 174

Preface

The goal of this project-oriented plant layout and material handling textbook is to provide students and practitioners with a resource which describes the techniques and procedures of plant layout and material handling. This is a "how-to" practical book which will lead the reader through the collection, analysis, and development of information to produce a quality plant layout. The information and techniques must be developed in a systematic manner, so jumping ahead is discouraged. Once the plant layout practitioner experiences the results of following the sequence of steps, they will enjoy the seeming magic of a well-thought out plan. It takes several experiences before a practitioner can comfortably forget about producing a layout until all the data has been collected and analyzed.

The mathematical knowledge requirements for this textbook is high school algebra. The manipulation of numbers is extremely important to a quality plant layout job, but nothing above the knowledge of algebra will be required.

Some experience with a *computer-aided design* program (CAD) will be helpful for a career in plant layout. CAD is the way of the future.

It is said that the better the plant layout, the more profitable the company. The plant layout and material handling procedure starts with collecting information from other departments. Chapter 2 describes the sources of information that start the project. Very briefly, the marketing department tells us how many units to produce every day, the product engineering department gives us blue prints and bills of materials, and management policy gives us inventory policy, investment policy, make or buy decisions, start date, and organization charts. Once we have the above information, the project is defined and we (the manufacturing engineering department) start developing basic data.

Chapter 3 describes how we develop route sheets, the sequence of operation, the time standard, assembly charts, assembly line balancing, and the number of machines. Chapter 4 analyzes material flow in order to insure proper placement of machines and departments to minimize costs. There are seven techniques discussed in the chapter.

Chapter 5 describes the activity relationship diagram. This diagram records the importance of relationships between departments, people, offices, and services. The activity relationship diagram is used to create a dimensionless block diagram (a general plan where each department will be in direct relationship to each other).

Once the sizes of departments have been determined, we can move on. Chapter 6 discusses workstation design, Chapter 7 covers auxiliary services-space requirements, Chapter 8 discusses employee services-space requirements, and Chapter 11 covers office layout space requirements.

The dimensionless block diagram is used as a guide to area allocation, as discussed in Chapter 12. The area allocation procedure results in an area allocation diagram which looks like a plant layout. At this point, a plot plan and detailed layout can be made. Chapter 13 discusses the many techniques of layout construction.

There are many other functions that require space. Some of these areas require as much space as the production department. The stores and warehouse departments are good examples of this. Good analysis and knowledge of design criteria can save much space and promote efficiency of both personnel and equipment. Other functions and spaces like receiving, shipping, lunchroom, bathrooms, first aide rooms, and offices need careful consideration by the plant layout technician. Each location and size will affect the operation efficiency. Chapters 7, 8, and 11 are dedicated to those subjects. Chapter 14 covers selling the layout via a project report and oral presentation. This is an important part of any project.

The final plan (layout) is only as good as the analysis backing it up. In no area of industrial management will a technician affect the operation more than in plant layout and material handling. This is also one of the most enjoyable projects an engineer, manager, or technologist will ever be involved in.

This book is designed to assist the student in doing a plant layout project. Students should pick a simple product (manufacturing at least 10, parts which require five operators each) and do everything required to produce a layout to make at least 1,000 units per eight hour shift. The final project should be a written report and an oral presentation.

ACKNOWLEDGMENTS

I would like to acknowledge my teachers of Facilities Design: Richard Muther and Jim Apple. They are a part of me. I would also like to thank my wife, Mary, for her love and encouragement.

A special thank you to J. Edward Gaughran of S. I. Handling Systems, Inc. for his outstanding contribution—especially to the cover.

About the Author

Fred E. Meyers is a Registered Professional Industrial Engineer, Associate Professor of Industrial Technology, Director of the College of Engineering and Technology Applied Research Center, and the 1988–1989 Outstanding Teacher of the College of Engineering and Technology at Southern Illinois University at Carbondale, Illinois. He has 14 years of industrial engineering and production management experience with such companies as Caterpillar Tractor Company, Mattel Toy Company, Boeing Aerospace Division, Ingersol-Rand's proto tool division, and Spalding's Golf Club division. Mr. Meyers has worked as an industrial consultant since joining Southern Illinois University in 1975. He has consulted about his expertise with over 50 companies in many different industries such as energy, oil, sporting goods, transportation, appliances, distribution, lumber and plywood, paper manufacturing, furniture, tooling, fiberglass, and office work. His range of assignments has given him the experience to write this book.

Fred E. Meyers has taught plant layout and material handling to over 50 classes and thousands of students. He has created over 100 new product, warehouse, and office layouts.

Chapter 1

Introduction to Plant Layout and Material Handling

THE IMPORTANCE OF PLANT LAYOUT AND MATERIAL HANDLING

Plant layout and material handling affects the productivity and the profitability of a company more than almost any other major corporate decision. The cost of the product and, therefore, the supply/demand ratio are directly affected by the plant layout. A plant layout project is one of the best jobs an industrial technologist or manufacturing engineer will ever have. The project manager, after receiving corporate approval, will be responsible for spending a great deal of money. The project manager will also be held responsible for the timely, cost effective achievement of the goals stated in the project report and cost budget. The responsibilities of a project manager approach those of a company's president and only project managers who achieve or beat the stated goals will be given bigger projects.

Plant layout is the organization of the company's physical facilities to promote the efficient use of equipment, material, people, and energy. Plant layout is a sub-part of a wider subject called facilities design. *Facilities design* includes plant location and building design, as well as plant layout and material handling. Plant location decisions are made at the very top corporate level, often for reasons not having much to do with operation efficiency or effectiveness. Site selection may be a better subject for a political science class than a facilities design class. Every country, state, county, and town has an economic development program to attract new industry. The financial incentives to attract a company to a specific location can be very attractive. Therefore, the decision to locate there is not an en-

1

gineering decision. Other reasons for locating plants at specific sites can be personal. The company's president is from that town, so that's where we're going to build.

Building design is an architectural job. Their expertise in building design and construction techniques is extremely important to the facilities design project, and it is not unusual that the architectural firm will report to the plant layout project manager.

Material handling is defined simply as moving material. Material handling has affected (positively) working people more than any other area of work design. Today, we can say that the physical drudgery has been eliminated from work by material handling equipment. Every expense in business must be cost-justified, and material handling equipment is no exception. The money to pay for material handling equipment must come from the saving of labor, material, or overhead costs, and these monies must be recovered in two years or less (50 percent ROI or higher).

New plant construction is one of the largest expenses a company will ever have, and the layout of that plant will affect the people for years to come. The cost of the product will be affected by the initial layout as well. Improvements on that layout will keep the company competitive.

It is said that if you improve the flow of material, you will automatically reduce production costs. The shorter the flow through the plant, the better. The cost reduction formula from motion and time study is also valuable with plant layout and material handling.

1. Eliminate steps in the process.
2. Combine steps in the process.
3. Change the sequence of process to reduce travel distances and time.
4. Simplify the operation by moving steps closer together and/or automate the movement between steps.

Material handling accounts for about 50 percent of all industrial injuries and from 40 to 80 percent of all operating costs. The cost of equipment is also high, but a proper return on investment can be had. Keep in mind that we eliminate many industrial problems with material handling equipment. In no area of industrial history has more improvement been made than by use of material handling equipment.

Material handling and plant layout are so intertwined that they cannot be separated. The material handling equipment choice will affect the plant layout design. In this book, we will talk about plant layout techniques first and then material handling. But we must never forget that they are inseparable in the real world.

THE GOALS OF PLANT LAYOUT AND MATERIAL HANDLING

A mission statement can best communicate the primary goal of a plant layout and material handling project. For example:

Build me a plant to produce 1,500 swing sets per shift.

Cost and quality are so important that they can be built into the mission statement, but as a separate goal we would state our supporting objectives like this:

1. Minimize unit cost.
2. Optimize quality.
3. Promote the effective use of:
    a. people;
    b. equipment;
    c. space; and
    d. energy.
4. Provide for:
    a. employee convenience;
    b. employee safety; and
    c. employee comfort.
5. Control project costs.
6. Achieve the production start date.
7. Achieve miscellaneous goals.

Your mission statement should be simple. The mission statement is used to keep you on track and to help in all project decisions. Our mission is to provide a specific number of quality units per period of time at the lowest possible cost. It is not to show off our advanced manufacturing knowledge or to have a showplace for our computers and robots. The mission statement is there to remind you:

1. The first sub-goal is to minimize unit cost of the prime objective. This means that every dollar expended in excess of the cheapest method of getting into production must be cost-justified. This *does not* mean we buy the cheapest machine because the most expensive machine may produce the lowest unit cost. When products are new, production volume may be low. Not much can be spent on advanced manufacturing technology, but we still need equipment. This is when we buy the cheapest ones available.

2. Quality is critical and difficult to measure; we all know that a near-perfect car is available, a Rolls Royce, but how many can we sell? We can make a better product if we buy better materials, machine closer tolerances, add additional options, and the like. But is there enough market for this high quality, high cost item?

Mass production was made possible by providing products that the masses could afford. This called for lowering the design, cost, and quality. Top management of the auto industry might state as a quality standard:

Let's design a utility automobile that will last 100,000 miles.

The designers will design every part with their goal in mind. They may state more clearly that 95 percent of the autos will last 100,000 miles or more. The average, therefore, would be higher, but any cost spent to create any one part of better quality will be money misspent. Quality and cost are the two primary competitive fronts. Controlling one without the other will lead to failure. We must constantly balance cost and quality. In plant layout and material handling we must consider quality in every phase.

3. Promoting the effective use of people, equipment, space, and energy is another way of saying "reduce costs." People, equipment, space, and energy are our company's resources. They are expensive and we want to use them effectively. Productivity is a measure of use and means output divided by input. To increase productivity, we need to increase output, reduce input or a combination of the two. The location of services like restrooms, locker rooms, cafeterias, tool cribs, and any other service will affect employee productivity and, therefore, the employees' utilization or effectiveness. It is said, "We can run pipe and wire, but you cannot run people." Providing convenient locations for services will increase productivity.

Equipment can be very expensive and the operating cost must be recovered by charging each part produced on that machine a portion of this cost. The more parts we can run on one machine, the lower the unit cost each part must carry. So to achieve our second supporting objective, namely, reduce cost, we must strive to get as much out of each machine as possible. Calculate how many machines are required in the beginning for maximum machine use. Remember, machine location, material flow, material handling, and work station design all affect equipment usage.

Space is also costly and we want to promote effective use of our space. Good work station layout procedures will include everything required to operate that work station, but no extra space. Normally we can do a good job of using floor space, but what about the other spaces?

a. Under the floor is a good place for utilities, basements, walkways, under floor conveyors for scrap, and tank storage.

b. Overhead—to the rafters is usable space. We can use this space for overhead conveyors, pallet racks, mezzanine shelf storage, balcony offices, pneumatic delivery systems, dryers, ovens, and so on.

c. Ceiling space under the roof (in the rafters) is space that can be used for utilities, heating and cooling, sprinkler systems, cat walks, and some storage.

d. On the roof space can be used for parking, weather testing, utility units, ovens, and so on. We want to promote the use of all the space in the plant. Do not forget any area.

Energy costs can be big money. Million dollar utility budgets are common. We can promote better energy dollar use by good layout techniques. Open dock doors can allow heating and cooling energy to escape. Placing hot equipment where energy can be contained could reduce energy requirements. A silly example would be running the air conditioning while having a fire in the fireplace. Electricity, gas, water, steam, oil, and telephone must all be used efficiently. The plant layout will greatly affect these costs.

4. Providing for the convenience, safety, and comfort of our employees' convenience has already been discussed, but keep in mind that besides being a productivity subject, its an employee relations factor. If we design plants with inconvenient employee services, we are telling the employees every day that we don't care about them. Drinking fountains, parking lot design and location, employee entrances, as well as restrooms and cafeterias must be convenient to all employees.

Safety of our employees is a moral and legal responsibility of the plant layout person. The weight of tools and products, the size of aisles, the design of work stations, and housekeeping all affect the safety of our employees. Every decision made in plant layout and materials handling design must include safety considerations and consequences. Material handling equipment has reduced the physical demands of work and, therefore, has improved industrial safety.

But material handling equipment itself can be dangerous. The industrial safety statistics indicate that 50 percent of all industrial accidents occur on shipping and receiving docks while handling material. We must continue our fight to reduce injury by every means possible.

Good housekeeping is having a place for everything and having everything in its place. The term "everything" is all inclusive—tools, materials, supplies, empty containers, scrap, waste, and so on. If the layout hasn't considered any one of these items, a housekeeping problem will result and this clutter is dangerous and costly.

"Comfort" is a term which communicates plush, costly surroundings, but in work station design it refers to working at the correct work height and standing or sitting alternatingly. We don't want to add fatigue to the operator. When the operator is on a break, we want to provide comfortable surroundings so they can recuperate and return to work refreshed and therefore productive.

5. Control project costs. The plant layout and material handling project must be costed before presenting the plan to management for approval. What top management approves is the "spending" of money. Once the project is approved, the project manager is authorized to spend the budgeted monies. Going back for more money is harmful for your career. Budgeting and then living with the budget

Figure 1-1 New Product Work Progress Report To Be Completed By

	1670	1810
1. Product Number	1670	1810
2. Product Name	Wizbang	Jumbo
3. Project Engineer	Meyers	Meyers
4. Rate Per Shift	1,500	1,750
5. Manufacturing Plan	3–1√	4–1
6. Material Handling Plan	3–1√	4–1
7. Time Standards Set	3–5√	4–5
8. Determine Number of:	3–6√	4–6
A. Fabrication Machines Needed	3–6√	4–6
B. Assembly and P.O. Stations Needed	3–6√	4–6
9. Issue Flow Chart	3–10√	4–10
10. Design Work Stations	3–10	4–10
11. Design Material Handling	3–10	4–10
12. Develop Budget	3–12	4–15
13. Layout Plan	3–14	4–15
14. Presentation	3–15	4–15
15. Write Work Orders to Build Stations	3–25	4–15
16. Issue Purchase Orders	3–15	4–15
17. Develop Quality Control Requirements	4–1	5–1
18. Tryout First Stations	4–1	5–1
19. Install Equipment	4–14	5–14
20. Write Work Methods Sheets	4–14	5–14
21. Run Production Pilot	4–15	5–15
22. Production Start	5–1	5–30
23. Recheck Everything		

√ Means complete   Date 3/11/xx
Items 10 and 11 Behind Schedule

are things successful managers and engineers learn to do very early in their careers.

6. Achieve the production start date. Very early in the life cycle of a new product, the production start date is set. The success of the project will depend on whether we get our product to the market on time. With the quantity of product we are designing the plant to produce and the production start date given to us, we must meet these goals. If we start late, we may not be able to make up for the lost production. This is especially true for seasonable products. In fact, if you miss the season, you miss the whole year. Fast moving consumer products companies, like toy companies, will set the production start date and schedule backwards to establish a product schedule. Figure 1-1 is such a schedule. It cannot be overstated that schedules *must* be met.

7. Miscellaneous goals include additional goals and objectives of plant layout and material handling. These should be added as you and your management decide something is important. For example, you may want to promote something like:

1. Minimizing inventory.

2. Just-in-Time manufacturing.

3. Lifting no more than one part at a time.

4. No bending to pick up parts.

5. First in, first out inventory.

Whatever you think is important and should be accomplished can be stated as a goal. Goals are something to be strived for. One hundred percent achievement is not necessary, but without the stated goal, we have almost no chance of achieving anything. Goals must be measurable so progress can be measured.

THE PLANT LAYOUT PROCEDURE

The quality of our plant layout plan (the blueprint) will depend upon how well we collected and analyzed the basic data. The blueprint is the final planning step and plant layout people seem to want to start here. This is like reading the last page of a book first. We must resist jumping into the layout before collecting and analyzing the basic data. Many engineering professors and industrial consulting firms are

Figure 1-2 The 24-Step Plant Layout Procedure

1. Determine what will be produced (Chapter 2)
2. Determine how many will be made (Chapter 2)
3. Determine what parts will be made or purchased complete (Chapter 2)
4. Determine how each part will be fabricated (Chapter 3)
5. Determine sequence of assembly (Chapter 3)
6. Set time standards for each operation (Chapter 3)
7. Determine the plant rate (Chapter 3)
8. Determine the number of machines needed (Chapter 3)
9. Balance assembly lines (Chapter 3)
10. Study the flow requirement (Chapter 4)
    a) String Diagram
    b) Multi-Column Process Chart
    c) Form-to-Chart
    d) Process Chart
    e) Flow Process Chart
    f) Flow Diagram
11. Determine activity relationships (Chapter 5)
12. Layout each work station (Chapter 6)
13. Identify needs for personal and plant services (Chapters 7 and 8)
14. Identify office needs (Chapter 11)
15. Develop total space requirements (Chapters 6, 7, 8, and 11)
16. Select material handling equipment (Chapters 9 and 10)
17. Allocate area (Chapter 12)
18. Develop plot plan and building shape (Chapter 13)
19. Construct master plan (Chapter 13)
20. Seek input and adjust (Chapter 14)
21. Seek approvals (Chapter 14)
22. Install (Chapter 14)
23. Start up (Chapter 14)
24. Followup (Chapter 14)

trying to develop a formula for plant layouts. So far, they have developed computer algorithms and simulations for parts of the analysis. So far, the best approach is to take a systematic approach, one step at a time, adding information at each step. When completed this way, the results appear like magic (a great plant layout results). A mature layout technician knows that a good layout is inevitable if the procedure is followed.

The plant layout procedure is a general plan of the project. Each step will include some techniques that won't be used in every situation. Skipping steps is permissible if considered and determined not be necessary. The following 24-step procedure (Figure 1-2) is the basic for the remainder of the book. If you are doing a layout project, you might try this list as an outline.

TYPES AND SOURCES OF PLANT LAYOUT PROJECTS

1. *New plants:* By far the most fun and the most influence can be had with a new plant project. The fewest restrictions and constraints are placed on the project.
2. *New product:* The company sets aside a corner of the plant for a new product. The new product must be incorporated into the flow of the rest of the plant, and some common equipment may be shared with old products.
3. *Design changes:* Product design changes are always being made to improve the cost and quality of the product. The layout may be affected and every design change should be reviewed by the plant layout technician.
4. *Cost reduction:* The plant layout technical may find a better layout that will produce more product with less manpower. There are many people and many areas of cost reduction that will affect the layout.

Every enterprise has material and peopel flows. Disney World's flow is people; hospitals have flows of patients, medical supplies, and food service; stores have flows of customers and merchandise; kitchens have flows of cooks and foods. Every area of human activity has a flow. If we study the flow, we can improve it by changing the layout. Opportunities are everywhere.

It is said that only death and taxes are for sure. I would like to add a third sure thing—a plant layout will change. Some industries are more changeable than others. For example, a toy company may have new products added to their product line every month. Plant layout work would be continuous in such a company. In a paper mill, the layout would change very little from year to year, so plant layout work would be a minor task, but plant layout projects are everywhere.

QUESTIONS

1. What is plant layout?
2. What is facilities design?

3. What is material handling?
4. What are the four questions to be asked about any cost?
5. Material handling accounts for
   a) What percent of injuries?
   b) Operating costs?
6. List 11 goals of plant layout and material handling.
7. What is a mission statement?
8. What two items on Figure 1-1 are behind schedule?
9. What is the value of a plant layout procedure?
10. What is the plant layout procedure?
11. What are the four types of plant layout projects?

Chapter 2

Sources of Information

The plant layout is dependent upon some basic information. Much of this basic information comes from other departments within the company. The larger the company, the less information the plant layout technologist actually produces. Some companies have several subdepartments within manufacturing engineering. For example, the processes section would establish the routing and machine assignments; tool design would design the fixtures and specify the specific tools; time standards application section would set the time standards; and all this information would be available to the plant layout section.

This chapter will discuss the sources of information outside the manufacturing department, while Chapter 3 will discuss the information required from within the manufacturing department. The plant layout technologist will always need to get the external information from someone else, but we may need to produce the manufacturing department's information ourselves.

There are three basic sources of information outside the manufacturing department:

1. Marketing;
2. Product design; and
3. Management policy.

THE MARKETING DEPARTMENT

The marketing department provides a research function which analyzes what the world wants and needs. Our company's marketing department searches out ways we can fill the customers wants and needs. Some of the information that marketing provides is:

1. Selling price;
2. Volume;
3. Seasonality; and
4. Replacement parts.

Determining the selling price is not an exclusive function of the marketing department. The industrial engineering organization may supply the cost data for pricing, but the selling price has a direct influence on the number of units we sell. Every customer makes a value analysis on each purchase. The lower the price, the more customers choose our product. Pricing is very complicated and marketing, production and finance departments are all a part of this decision, but marketing needs that information before it can ask the customers "how many do you want to buy?"

Volume is reduced down to how many units do we want to build per day. The marketing department may take some model shop samples out to a few important customers and ask them their opinions. If these customers like the new product, they would tell us that they would buy so many. Using past experience, we know that 20 percent of our customers buy 80 percent of our total production. Therefore, if a small group of customers say they will buy 125,000 units and they represent 50 percent of our annual sales, 250,000 units will be needed. If we work 250 days a year (50 weeks times 5 days per week), then 1,000 units would be needed every day.

The number of units required per day is very important number for the plant layout project. The plant rate is calculated from this number.

Example: We need 1,000 units per shift out of our plant. Four hundred eighty minutes are available in an eight-hour shift (8 hrs. × 60 min./hr.). Subtract 10 percent for down time (personal fatigue and delay time). This leaves 432 minutes left to produce 1,000 units. But neither people nor plants work at 100 percent. They work from 60 to 120 percent depending on the quality of management. Let's say our plant will work at 90 percent efficiency, then 432 min. × 90 percent = 389 effective minutes of work can be expected from each person in the plant.

$$R = \frac{389 \text{ min.}}{1,000 \text{ units}} = .389 \text{ min./unit}$$

R = Plant Rate = .389 min. per unit

CHAP. 2: SOURCES OF INFORMATION

A finished, packed out unit must come off the assembly and packout line every .389 minutes or just about 2.5 units per minute. That means every work station and every machine in the plant will need to produce about 2.5 parts per minute.

$$\frac{1 \text{ min}}{.389 \text{ min. per unit}} = 2.57 \text{ units/min.}$$

If we need two parts (like axles on a wagon) per finished unit, then we would need 5.14 parts per minute.

The plant rate is one of the most important numbers in plant layout. We will use it in the next chapter to calculate number of machines, work stations and employees required by our plant.

Seasonality is important. Customers want space heaters and sleds in the winter, gas grills and swimming pools in the spring and summer, and toys must be in the stores for Christmas. If we waited until just before the season to start producing our product, we would either need a great deal of extra machines or we would miss our market time. If we produce all year long just for the Christmas season, we would need 10 to 12 months of warehouse space. Determining how soon to start and how much to make per day is a compromise between inventory carrying cost and production capacity cost. The objective is to minimize total cost.

The subject of production and inventory control goes hand in hand with plant layout, and production and inventory control policies will have a big affect on plant layout.

Recognize the need for replacement parts. If you have been in business for any period of time, your product will begin to wear out. Customers may call you for replacement parts that have worn out or have broken. This business not only requires you to build extra inventory, but you must have storage and shipping areas for such customer service.

THE PRODUCT DESIGN DEPARTMENT

Blueprints, bill of materials, assembly drawings, and model shop samples tell the plant layout person about the prime mission. The product design department is the source of this valuable information. The first question anyone would ask when assigned a new layout project is "What are we going to make?" The output of the product design department tells us exactly what we are going to manufacture.

Blueprints, sketches, pictures, CAD drawings and samples all communicate the idea of what we want to build. (See Figures 2-1, 2-2, and 2-3.) There will be drawings of each individual part of the product as in Figure 2-1. These drawings will tell us size, shape, material, tolerances, and finish. Assembly drawings (see Figure 2-2) show many parts (if not all parts) and how they fit together. An exploded drawing (Figure 2-3) is an especially useful drawing. It shows how parts

Figure 2-1a Sample Blueprint for Tool Box Handle

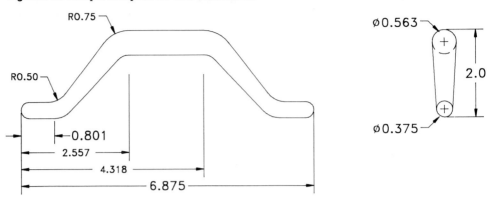

Figure 2-1b Part Drawing—Clamp and Plate

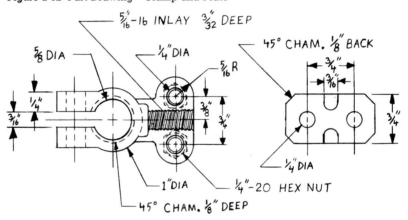

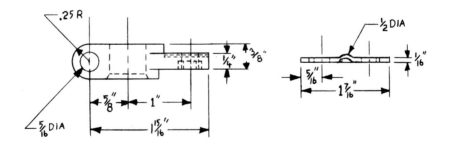

CHAP. 2: SOURCES OF INFORMATION

Figure 2-2a Sample Sketch for Tool Box and Tray

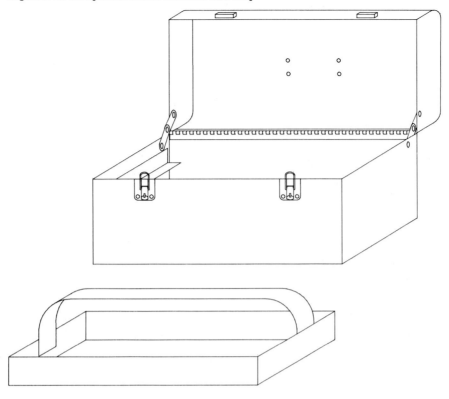

Figure 2-2b Sample Sketch, 3-D View

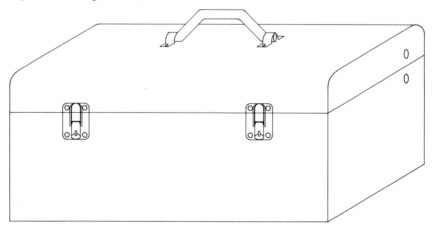

Figure 2-3a Sample Sketch—Exploded View

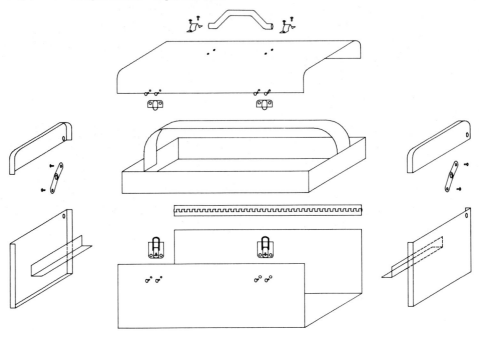

Figure 2-3b Exploded Drawing—Battery Cable Clamp

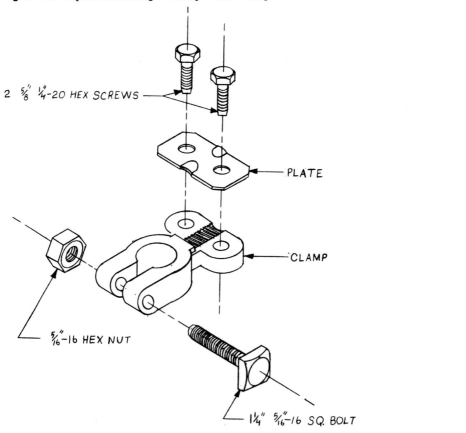

2 $\frac{5}{8}''$ $\frac{1}{4}''$-20 HEX SCREWS

PLATE

CLAMP

$\frac{5}{16}''$-16 HEX NUT

$1\frac{1}{4}''$ $\frac{5}{16}''$-16 SQ. BOLT

fit together using centerlines and separated parts aligned to indicate the sequence of assembly. When the layout person is working on the assembly line layout, the exploded drawing will be the guide. The plant layout could not get started without blueprints or sketches.

Either a *parts list* or a *bill of materials* will be provided to the layout technologist by the product engineering department with each new product. (See Figure 2-4.) Parts list and bill of materials are the same thing and list all the parts that make up a finished product. Parts list include part numbers, part names, the quantity of each part, what parts make up sub-assemblies, and may include material specifications, parts and raw material unit costs and make or buy decisions. The make or buy decisions are a total management decision, not just product engineering department, but the parts list is a good place to indicate that decision.

Figure 2-4 Tool Box Parts List

Part #	Part Name	Quantity Needed Per Unit
1	Body Ends	2
2	Tray Bracket	2
SSSA1	Body End	2
3	Body	1
SSA1	Body Assembly	1
4	Cover End	2
5	Cover	1
SSA2	Cover Assembly	1
6	Hinge	18"
SA1	Tool Box	1
7	Tray Ends	2
8	Tray Body	1
9	Tray Handle	1
SA2	Tray	1
10	Paint	AR
11	Handle	1
12	Clips	2
13	Rivet	4
A1	Tool Box	
14	Catch	2
15	Strike	2
16	Rivets	8
A2	Tool Box	
17	Hinges	2
18	Rivets	4
A3	Tool Box	
19	Packing List	1
20	Registration Card	1
21	Name Tag	1
22	Dividers	4
23	Plastic Bag	
SA3	Parts Bag	1
24	Carton	1
25	Tape	24"
Final Packout		1

Companies themselves do not fabricate every part of its product. The parts that are purchased complete are called buyouts and can be fabricated cheaper by someone else. Some companies purchase every part complete from outside. These are called assembly plants. Those parts that we "make" are basic requirement for the fabrication end of our plant layout.

Model shop samples are hand made, very costly, exact models of what the product engineering department wants produced. They are not always available but, if they are, they are very helpful. Being able to "feel" the parts, to take them apart and reassemble them, to look at each part and consider how to make that part, are all easier to do with actual parts. Our understanding of the product is faster and more complete when using a model.

A model shop sample would be used as follows:

1. Unpack the unit noting the sequence of unpacking. This will be the basic information for our packout line. Be sure to keep good notes; photographs are very useful.

2. Play with the finished product to see how it functions. A good understanding of the finished unit's purpose is very helpful.

3. Disassemble the product very carefully. Keep good notes again. Then, reassemble the product. This will be your basic information about the assembly line.

4. Disassemble and look at each part. Decide which parts we are going to "make" in our plant and which parts are going to be purchased complete (buyouts).

5. Those parts that are going to be made by us in our plant will need further studies to determine how they will be fabricated from raw material. This is the subject of the next chapter.

Without a model shop sample, we would still need to do the above, except from drawings, prints, or sketches. The model shop sample makes the process easier and the results are better.

The product engineering department can be very helpful to the plant layout technicians. They can point out special manufacturing problems, critical relationships, dimensions and functions. The product engineer and the plant layout designer need to work together.

MANAGEMENT POLICY INFORMATION

Management refers to upper level employees who are responsible for the financial performance of a company. Such information like:

1. inventory policy;
2. investment policy;

3. start-up schedules;
4. make or buy decisions;
5. organizational relationships; and
6. feasibility studies

will have an affect on the plant layout. Layout designers must understand these policies up front or they may waste a lot of time.

Inventory Policy

The company's inventory policy could be something simple like "provide space for a one month supply of raw materials, work in process and finished goods." Those inventories would require space and facilities, but once the quantity to be stored has been determined, calculating the space requirement is easy. Stores and warehouse layouts will be discussed in Chapter 8. *Just in Time* (JIT) is a concept which reduces inventory and, therefore, space. JIT is an inventory policy and it will affect our layout. *Work in process* (WIP) needs space.

Investment Policy

The corporate investment policy is communicated in terms of *return on investment* (R/I). Return is the savings, and investment is the cost. If a project saves a large enough percentage of the cost, then it is a good idea. For example, a plant layout project might be approved with at least a 33 percent return on investment. Thirty-three percent is also a three-year pay-back period. Plant layout projects are one of the few investments that management will allow such a long pay-back period. Most cost reduction work requires a return on investment of greater than 100 percent or a pay-back period of less than one year.

When presenting a plant layout proposal to management for their approval, what you are seeking is an approval to spend so much money. The project engineer must collect costs from suppliers, vendors, maintenance personnel and the like, and then prepare a budget. We discussed budgets earlier, but remember the importance of staying under budget.

Start-up Schedule

We are given a project to start up a new product. We are usually told something like this:

> Provide a production facility to manufacture 1,200 gas grills per day beginning on November 15 of this year.

All the work needed to achieve the above will be backdated from November 15. For example:

	Completed By
Production start	November 15
Install equipment	November 1
Approval and order equipment	October 1
Master plan complete	September 15
Develop plot plan and allocate area	September 1
Select M.H. equipment	August 25
Develop total space requirement	August 20
Layout work station	August 15
Determine activity relationships	August 10
Identify office needs	August 5
Identify personnel and plant service needs	August 1
Develop flow requirements	July 25
Balance assembly lines	July 15
Determine number of machines	July 15
Set standards and plant rate	July 10
Determine assembly sequence	July 6
Develop route sheets	July 4
Make or buy decision	July 2
Determine what and how many will be made	July 1

Make or Buy Decisions

Do we make this part (fabricate from raw materials) or do we buy this part complete from a supplier who specializes in this kind of part? (See Figure 2-5.) The decision is normally quite straightforward and easy. If we are an existing company with a product line, we know what we can make and what we can't. If we are a new company, we may buyout all the parts and become an assembly operation only. As we get going, we may start making a few parts ourselves. No plant would make their own nuts, bolts, screws, tires, gauges, engines, bearings, tapes, and the like; someone has special equipment that makes them faster than we could and at a lower cost than we could ever achieve. The fabrication section of our manufacturing department is always in competition with the purchasing department of our own company because the cheapest way to provide the part to the assembly department is the best source.

The make parts are the subject of our fabrication layout. If we make no parts, no layout is needed. If we make a lot of parts, a large layout project results.

Organizational Relationships

An organization chart communicates much to the plant layout designer. (See Figure 2-6.) The number of employees determines the size of many areas like cafeterias, restrooms, office space, and medical facilities. The relationships among the various functions determines the closeness requirement for each department to every other department.

Figure 2-5 Tool Box Make or Buy Decision

Part #	Part Name	Quantity Needed Per Unit	Make or Buy
1	Body Ends	2	M
2	Tray Bracket	2	M
SSSA1	Body End	2	
3	Body	1	M
SSA1	Body Assembly	1	
4	Cover End	2	M
5	Cover	1	M
SSA2	Cover Assembly	1	
6	Hinge	18″	M
SA1	Tool Box	1	
7	Tray Ends	2	M
8	Tray Body	1	M
9	Tray Handle	1	M
SA2	Tray	1	
10	Paint	AR	B
11	Handle	1	B
12	Clips	2	B
13	Rivet	4	B
A1	Tool Box		
14	Catch	2	B
15	Strike	2	B
16	Rivets	8	B
A2	Tool Box		
17	Hinges	2	B
18	Rivets	4	B
A3	Tool Box		
19	Packing List	1	B
20	Registration Card	1	B
21	Name Tag	1	B
22	Dividers	4	B
23	Plastic Bag	1	B
SA3	Parts Bag	1	
24	Carton	1	B
25	Tape	24″	B
Final Packout		1	

Feasibility Studies

Many new product ideas are recommended to management. These ideas need to be evaluated before they are accepted as new plant layout projects. The early stages of a plant layout project might be referred to as a feasibility study. Feasibility studies are generally for future projects and highest level project managers and engineers perform these studies. Out of four feasibility studies, we may get one project.

Figure 2-6 Tool Box Manufacturing Plant Organizational Chart

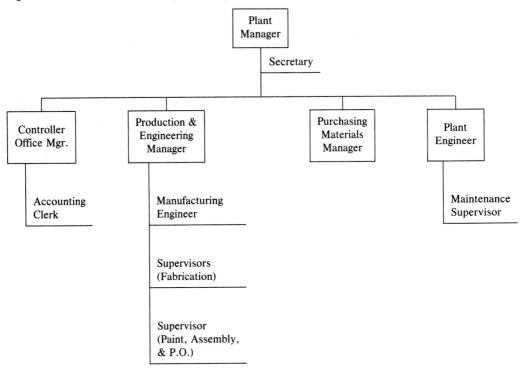

CONCLUSION

The product engineering design department gives us blueprints and bill of materials to help us understand what is going to be manufactured in our plant. These blue prints and bill of materials will help decide what parts will be made inside our plant, and what parts will be purchased from outside sources. Those parts that are made (manufactured) inside our plant will require manufacturing plans as discussed in Chapter 3.

The marketing department researches the potential market demand for our new or redesigned product and provides us with a quantity to produce per period of time. We will always break this quantity down into units per day to determine number of machines and number of people needed.

Management policy communicates the company's attitudes and decisions. The return on investment, the inventory policy, the start up date required all have significant affects upon our plant layout and material handling project.

A plant layout and material handling project will be held up until the information provided by other departments is provided. We are not the experts in product design, marketing or management policy. We should not make these decisions.

QUESTIONS

1. What are the three broad sources of information?
2. What information does marketing provide?
3. What is an *R* value?
4. What is included in an *R* value calculation?
5. Why is the *R* value so important?
6. What information do we get from the product design department?
7. What information do we get from management policy?
8. What is a make or buy decision?
9. Who is the competition for our fabrication department? Why?

Chapter **3**

Process Design

How are we going to make each part? What is the equipment, tools and time standards required to make each part? What is the sequence of assembly? How fast do the conveyors need to run? How do we divide the work evenly among the assemblers and packers? These are the kinds of questions that must be answered by the process designers. The process designer and the plant layout designer may be the same person but in large companies, they are separate departments.

Process design can be divided into two broad categories, *fabrication* and *assembly*. Fabrication process design is initially planned on a route sheet. Assembly and packout process design uses the techniques of assembly charts, and assembly line balancing.

FABRICATION: MAKING THE INDIVIDUAL PART

The sequence of steps required to produce (manufacture) a single part is referred to as the *routing*. We route the part from the first machine to the second machine and so on until we have a finished part that will be united with other parts. The form used to describe this routing is called the *route sheet*.

Route Sheets

A route sheet (see Figure 3-1) is required for each individual fabricated part (make part). If the product to be manufactured has 30 parts and we buy 10 and make 20 ourselves, 20 route sheets will be required. The route sheet lists the operations

Figure 3-1 Sample Route Sheets

Route Sheet 1

PART NUMBER_____7440_____   PART NAME____Body____

RAW MATERIAL_____1,020 Cold Rolled Steel 18″ × 24″ 20 ga

ORDER QUANTITY_____

Operation #	Machine Name	Operation	Pieces Per Hour Time Standard
5	Strip Shear	Cut to Width	1,400
10	Chop Shear	Cut to Length	1,175
15	Punch Press	Punch Catch Holes	650
20	Press Brake	Form Two Legs	475

Route Sheet 2

PART NUMBER_____7420_____   PART NAME____End____

RAW MATERIAL_____1,020 Cold Rolled Steel 6 1/2″ × 6 1/4″ 20 ga

ORDER QUANTITY__Twice the Number of Bodies__

Operation #	Machine Name	Operation	Pieces Per Hour Time Standard
5	Strip Shear	Cut to Width	1,850
10	Chop Shear	Cut to Length	950
15	Punch Press	Punch Hinge Holes	825
20	Press Brake	Form Three Sides	595

There are 7 more sheets for our tool box pictured in Figure 2-2. These include:

Cover ends;

Cover body;

Tray basket;

Tray ends;

Tray body;

Tray handle; and

Hinge.

required to make that part in the proper sequence. The route sheet gets its name from the way its used. For example, we need to produce an order of 2,500 axles for a wagon. A copy of the route sheet would be issued by the production and inventory control department showing the order quantity. This order would then be given to the stores department where the raw material for 2,500 axles would be pulled and shipped to the first operation (according to the route sheet). The route sheet would accompany the material from operation to operation telling the operators what to do. The route sheet will tell the plant personnel about the:

Part number;

Part name;

Quantity to produce (left blank until needed);

Operation number;

Operation description;

Machine number (if available);

Machine name;

Tooling needed; and

Time standard.

The route sheet ends with the last operation prior to being assembled with other parts. For example, if parts are going to be welded together, the individual parts lose their identity once joined with another part, so the route sheet would end before welding. If an individual part goes through a clean, paint, and bake operation before being assembled, then the clean, paint, and bake procedure would be included on the route sheet.

The sequence of operations as shown on the route sheet affects the proper layout of the equipment on the plant floor. We want the material to flow smoothly through the plant from the first machine next to raw material stores to the very next machine right next to the first. This will ensure that the part travels as short a distance as possible. Skipping over machines and backtracking must be discouraged. When many parts are fabricated in one group of machines, jumping around may be necessary, but we want to minimize this jumping, skipping and backtracking. There are two ways to change the sequence in order to make the flow through the plant smoother:

1. Change the route sheet if possible so that the sequence of operation agrees with the other parts or the plant layout.
2. Change the layout so that the machines are in the correct sequence.

Changing the paperwork is our first choice because its the cheapest way.

The time standards are an important part of the route sheets. Time standards are used to determine how many machines are needed in our layout. Time standards are another piece of information that may come from another group, but in many companies time standards are developed by the plant layout designer.

Figure 2-3 is an exploded drawing of a tool box. This tool box has nine different manufactured parts:

Part #	Part Name	Quantity Per Box
1	Body Ends	2
2	Shelf Brackets	2
3	Body	1
4	Cover Ends	2
5	Cover Body	1
6	Hinge	1
7	Tray Ends	2
8	Tray Body	1
9	Tray Handle	1

Nine route sheets will be needed. Figure 3-1 shows two of these route sheets. Figure 3-2 is a summary of all nine route sheets.

The Number of Machines

How many machines do we buy? This question can only be answered when we know:

1. How many finished units are needed per day?
2. Which machine runs what parts?
3. What is the time standard for each operation?

How many finished units are needed per day? The marketing department tells us how many products to produce (manufacture) per day. For example, let's say we are going to build 2,000 tool boxes per eight-hour shift. From the figure we calculated the plant rate (R value) as follows:

60 min./hr. × 8 hrs.	= 480	minutes
Less Down Time 10%	− 48	minutes
Available Minutes/Shift	= 432	minutes
Expected Efficiency @	80%	
Effective Minutes	= 345.6 minutes	

We have 345.6 minutes to produce 2,000 units.

$$R \text{ Value} = \frac{345.6 \text{ min.}}{2,000 \text{ units}} = .173 \text{ minutes/unit.}$$

Which machine runs what parts? The route sheets produced in the previous section will tell us which machines are needed to produce each part. Figure 3-2 is a summary of the nine route sheets needed to manufacture the tool box. Note from Figure 3-2 that seven different parts are run on the strip shear, eight different parts are run on the chop shear, four different parts are run on punches, seven parts on the press brakes and only one part on the roll former.

What is the time standard for each operation for each part? The time standards for every operation on every part is in both pieces per hour and decimal minute (see Figure 3-2). We need the decimal minute time standards to compare with the R value calculated in question 1 of the section.

Once we know the plant rate (R value), the machines and the time standards, we divide the time standard (decimal minute) by the R value. The resultant number of machines should be in two decimal places (i.e., .34 machines). Once all the machine requirements for each operation have been calculated, we total the like

Figure 3-2 Summary of Route Sheets

Part Name	Body Ends	Tray Brackets	Body	Cover Ends	Cover	Hinge	Tray Ends	Tray	Handle
			Time Standards in Pieces Per House						
Parts/Unit	2	2	1	2	1	1	2	1	1
Operations									
Strip Shear	1,850	2,750	1,400	2,100	1,750	—	2,250	1,850	—
Chop Shear	950	1,400	1,175	1,050	1,320	935	1,220	1,410	—
Punch	825	—	650	870	759	—	—	—	—
Form	595	841	475	616	528	—	629	567	—
Roll Form	—	—	—	—	—	—	—	—	375
			Time Standards in Decimal Minutes/Unit (Divide The Above Pieces Per Hour Into 54 Minutes)						
Strip Shear	.029	.020	.039	.026	.031	—	.024	.029	—
Chop Shear	.057	.039	.046	.051	.041	.058	.044	.038	—
Punch	.065	—	.083	.062	.071	—	—	—	—
Form	.091	.064	.114	.088	.102	—	.086	.095	—
Roll Form	—	—	—	—	—	—	—	—	.144

The time standards in decimal minutes will be used to determine the number of machines needed.

machines and round up recommending to purchase enough machines. Figure 3-3 is the result of dividing Figure 3-2 by the R value of .173 (so we can produce 2,000 tool boxes per shift). This information on the number of machines required will be used later to determine the number of square feet of floor space needed in our fabrication department. Chapter 6 is workplace design and space determination, and that space is based on the above fabrication equipment requirements.

We will need to acquire two strip shears, four chop shears, three punch presses, six press brakes and one roll former to produce 2,000 tool boxes per day. Always round up on the total machines otherwise a bottleneck will be created and 2,000 tool boxes per day won't get built unless our plant works overtime. A very expensive machine could cause us to round down, but that machine will be a continuing problem forever. I always recommend rounding up.

Figure 3-3 Machine Requirements Spreadsheet

Part Name	2 Body Ends	2 Tray Brackets	1 Body	1 Cover Ends	1 Cover Body	1 Hinge	2 Tray Ends	1 Tray Body	1 Tray Handle	Total Machines
Machines										
Strip Shear	.34*	.24	.23	.30	.18	—	.28	.17	—	1.74
Chop Shear	.66	.45	.27	.59	.24	.33	.51	.22	—	3.27
Punch Press	.75	—	.48	.72	.41	—	—	—	—	2.36
Press Brake	1.05	.74	.65	1.02	.59	—	.99	.55	—	5.59
Roll Former	—	—	—	—	—	—	—	—	.83	.83

* .029 ÷ .173 ×2 parts/unit = .34 machines.

Once all parts are produced by the fabrication departments or received from the suppliers and available for assembly, new analytical tools are needed. Subassembly, welding, painting final assembly, and packout are all functions included in this area of the plant.

The Assembly Chart

The *assembly chart* (see Figure 3-4) shows the sequence of operations in putting the product together. Using the exploded drawing (Figure 2-3a) and the parts list (Figure 2-4), the layout designer will diagram the assembly process. The sequence of assembly may have several alternatives. Time standards are required to decide which sequence is best. This process is known as *assembly line balancing*.

Time Standards are Required for Every Task

The tasks should be as small as possible so the layout designer has the flexibility of giving that small task to several different assemblers (see Figure 3-5a). The time

Figure 3-4 Assembly Chart—Tool Box

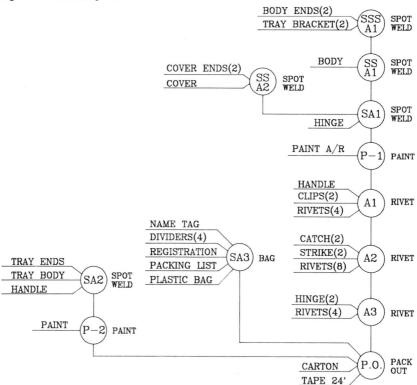

　　　　　　　　　　　　　　　　　　　　CHAP. 3: PROCESS DESIGN

Figure 3-5a Assembly and Packout Time Standards

Operation #	Operation Description	Minutes
SSSA1	Assemble and spotweld two tray brackets to two body ends	.153 each
SSA1	Weld two body ends to body	.291
SSA2	Weld two cover ends to cover body	.260
SA1	Weld hinge to body, weld cover to hinge, and hang an overhead conveyer	.356
P1	Clean, paint, and bake	automatic
A1	Rivet handle and 2 clips to cover	.310
A2	Rivet 2 catches to cover 2 strikes to body	.555
A3	Rivet 2 hinges to body and cover	.250
SA2	Weld tray ends to tray and add handle. Hang on overhead conveyor	.415
P2	Clean, paint, and bake tray	automatic
SA3	Bag loose parts	.250
P.O.	Pack tray into tool box, place plastic bag into tool box and close. Form carton and pack tool box	.501

standard setting techniques used for assembly line design are from either a predetermined time standard system or standard data. If we had 10 bolts to assemble, we would want a time standard (decimal minute) per bolt and a separate time for running the bolt down and tightening because this process gives us the most flexibility.

Plant Rate and Conveyer Speed

Conveyer speed is dependent on the number and units needed per minute, the size of the unit, the space between units, and, sometimes, the hook spacing. *Conveyer belt speed* is recorded in feet per minute. Therefore, the size of the part plus the space between parts (measured in feet) times the number of parts needed in one minute equals feet per minute.

Example: Charcoal grills are in cartons $30'' \times 30'' \times 24''$ high. A total of 2,400 grills are required every day.

	480	minutes per eight-hour shift
less	50	minutes scheduled down time (breaks, etc.)
	430	available minutes
	80%	anticipated performance
	344	effective minutes of work per day
÷	2,400	grills per day
=	.143	minutes per grill
	1	minute = 7 grills/minute
	.143	min/grill

7 grills per minute $\times$ 2-1/2 ft./grill $(30'')$ = 17.5 ft./min.

ASSEMBLY AND PACKOUT PROCESS ANALYSIS

Our conveyer must run at 17.5 ft./min. or we will not produce our 2,400 grills. Check of production rate:

$$
\begin{array}{rl}
7 & \text{grills per minute} \quad \times \quad 430 \text{ min./shift} \\
\times \quad 80\% & \text{performance} \quad = \quad 2,408 \text{ grills/shift.}
\end{array}
$$

Paint Conveyer Speed

The overhead paint conveyer speed is complicated additionally by multiple parts per hook and hook spacing.

Example: We are going to paint the following parts on one overhead conveyer system:

Part Number	Parts/Hook	Quantity To Paint (per shift)	Needed Hooks/Day	Hooks Per Minute*
15	1	500	500	1.45
263	4	300	75	.22
44	2	1,000	500	1.45
14	8	2,000	250	.73
21	2	100	50	.15
03	1	125	125	.36
				4.36

* Based on 430 minutes @ 80% = 344.

A total of 4.36 hooks need to pass any one point in one minute if the hooks are one foot apart. So, 4.36 ft./min. would be the conveyer speed. If the hooks were 1-1/2' apart, the conveyer speed would be 6.54 ft./min. (4.36 × 1.5). Figure 3-5b shows some different kinds of paint hooks.

This conveyer speed also determines the size of your drying oven and baking oven. Let's say that your parts need 10 minutes at 400° to dry. Ten minutes at 6.54 ft./min. equals 65.4' of conveyer in an oven. Most drying ovens take the parts into and out of the oven from the same end, so our oven will be about 33' long and 4-1/2' wide.

Let's look at the tool box example. We will use hook spacings of 18" because of our part size. Which parts need painting? The box and tray assembly only. Each is placed on a hook all by itself and sent through an electrostatic spray system. But first, it must be cleaned and dried. After painting, it will be baked and cooled. The drying needs 450° for 10 minutes. Fifteen minutes is required for cool down before an assembler can take it off the line.

	Parts/Hook	Quantity Per Shift	Needed Hooks/Day	Hooks Per Minute*
1. Box Assembly	1	2,000	2,000	5.78
2. Tray Assembly	1	2,000	2,000	5.78
				11.56

* Based on our *R* value of .173 or 5.78/min.

CHAP. 3: PROCESS DESIGN

Figure 3-5b Paint Hooks

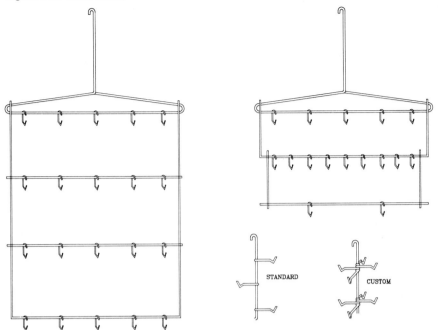

STANDARD   CUSTOM

A total of 11.56 hooks are needed every minute and each hook is 18″ apart (1-1/2′), so 11.56 × 1.5 = 17.34 ft./min. conveyer speed.

We need 10 minutes of drying and baking time, so 17.34 ft./min. × 10 min. = 173 feet of drying and baking. How about a double decker dryer 90 feet long with drying on the top (hottest) and baking on the bottom. Cooling time of 15 minutes × 17.34 ft./min. means 260 feet from baking to the first assembler.

Do you see the information is building and the data for your layout is coming forth? It is like magic, but we have taken the magic out and substituted a systematic approach.

When the tool boxes are taken off the overhead conveyer line after painting to rivet on the handle, the finished tool box can be placed on a flat belt conveyer. That conveyer speed needs to be 5.78 tool boxes per minute times the spacing. Since this tool box is 18″ long, 24″ between centers should be good spacing. 2 ft./box × 5.78 boxes/min. = 11.56 ft./min. The tray will stay on the hook until it reaches the packout station. Trays will be on every other hook, tool boxes on the other hooks.

We really have enough information to layout the clean paint and bake area now, but we will wait until Chapter 6 and calculate all production space requirements then.

Assembly Line Balancing

The purpose of the assembly line balancing technique is to:

1. Equalize work load among the assemblers.
2. Identify the bottleneck operation.

3. Establish the speed of the assembly line.
4. Determine the number of workstations.
5. Determine the labor cost of assembly and packout.
6. Establish the percentage workload of each operator.
7. Assist in plant layout.
8. Reduce production cost.

The assembly line balancing technique builds on the assembly chart (Figure 3-4) time standards (Figure 3-5) and the plant rate (R value) calculated in the last section.

The objective of assembly line balancing is to give each operator as close to the same amount of work as possible. This can only be accomplished by breaking the tasks into the basic motions required to do every single piece of work and reassembling the tasks into jobs of near equal time value. The workstation or stations with the largest time requirement is designated as the 100 percent station and limits the output of the assembly line. If an industrial technologist wants to improve the assembly line (reduce costs), he or she would concentrate on the 100 percent station. Reduce the 100 percent station in our example by 1 percent and save the equivalent of .25 people, a multiplying factor of 25 to 1.

An example assembly balance problem for our tool box example appears in Figure 3-6.

SA3 could be taken off the assembly line and handled completely separate from the main line and we can save money. SA3 .250 = 240 pieces per hour and .00417 hours each. If balanced the standard would be 180 pieces per hour and .00557 hours each.

	.0057	balanced cost
−	.00417	by itself cost
	.00140	savings hours/unit
×	500,000	units/year
	700	hours per year
@ $	7.50	per hour
=	$5,250.00	per year savings

This is called the *cost of balancing*. In this case its too high.
Sub-assemblies that can be taken off the line must be:

1. Poorly loaded. The less percent loaded the more desirable to be sub-assembled. For example, a 60 percent load on the assembly line balance would indicate 40 percent lost time. If we take this job off the assembly line (not tied to the other operators) we could save 40 percent of the cost.

2. Small parts easily stacked and stored.

3. Easily moved. The cost of transportation and the inventory cost will go up, but because of better labor utilization, total cost must go down.

Figure 3-6 Initial Assembly Line Balance

	Time Standard	Number Stations	Rounded Off	Average Time	Percent Loaded	Hours Per Unit	Pieces Per Hour
SSSA1	.306	1.77	2	.153	92	.00557	170
SSA1	.291	1.68	2	.146	87	.00557	180
SSA2	.260	1.50	2	.130	78	.00557	180
SA1	.356	2.06	3	.119	71	.00834	120
A1	.310	1.79	2	.155	93	.00557	180
A2	.555	3.20	4	.139	83	.01112	90
A3	.250	1.44	2	.125	75	.00557	180
SA2	.415	2.40	3	.138	83	.00834	120
SA3	.250	1.44	2	.125	75	.00557	180
P.O.	.501	2.90	3	.167	100	.00834	120
TOTAL	3.494		25			.06950	

A sub-assembly on the assembly line balance form would look like this:

	Time Standard	Number Stations	Rounded Off	Average Time	Percent Loaded	Hours Per Unit	Units Per Hour
SA3	.250	1.44	1.44	2.50	Sub	.00417	240

Look back at Figure 3-6 and SA3. We have saved plenty, but can we do this to SA1? No! Because it's a large part not easily stacked, stored, or moved.

The assembly line balance in Figure 3-6 is not a good balance because of the low percentage loads. An improvement is possible (look at the 100 percent station). If we add a third packer, we will eliminate the 100 percent station at P.O. Now the new 100 percent (bottleneck station) is A2 (93 percent). By adding this person, we will save 7 percent of 25 people or 1.75 people and increase the percent load of everyone on the assembly line (except P.O.). We might now combine A1 and A2, and further reduce the 100 percent. The best answer to an assembly line balance problem is the lowest total number of hours per unit. If we add an additional person, their time is in the total hours. Try to improve the tool box assembly line balance, then see how that affects the assembly lines in Figures 3-7 and 3-8.

Notes on the assembly line balance (Figure 3-6).

1. The busiest workstation is P.O. It has .167 minutes of work to do per packer. The next closest station is A1 with .155 minutes of work. As soon as we identify the busiest workstation, we identify it as the 100 percent station, and communicate that this time standard is the only time standard used on this line from now on. Every other workstation is limited to 360 pieces per hour. Even though other workstations could work faster, the 100 percent station limits the output of the whole assembly line.

2. .06950 hours is the total hours required to assemble one finished tool box. The average hourly wage rate times .06950 hours per unit gives us the assembly and packout labor cost. Again, the lower this cost the better the line balance.

Figure 3-7 Spot Weld Sub-Assembly

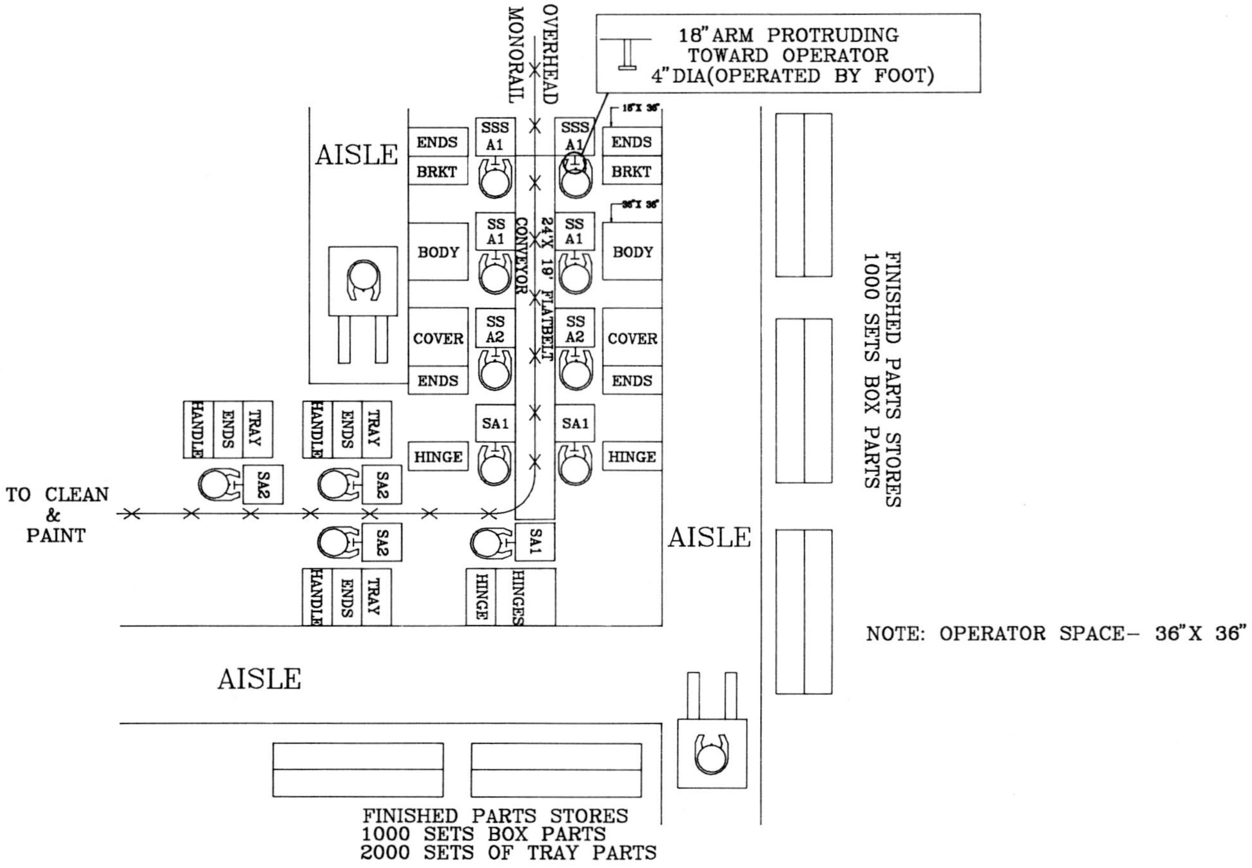

NOTE: OPERATOR SPACE— 36"X 36"

18" ARM PROTRUDING TOWARD OPERATOR 4" DIA(OPERATED BY FOOT)

FINISHED PARTS STORES
1000 SETS BOX PARTS

AISLE

ENDS · BRKT · BODY · COVER · ENDS · HINGE

SSS A1 · SS A1 · SS A2 · SA1

OVERHEAD MONORAIL

24'X 19' FLATBELT CONVEYOR

SSS A1 · SS A1 · SS A2 · SA1

ENDS · BRKT · BODY · COVER · ENDS · HINGE

SA1 · HINGES · HINGE

AISLE

TRAY ENDS HANDLE · SA2 · SA2 · TRAY ENDS HANDLE

TRAY ENDS HANDLE · SA2

TO CLEAN & PAINT

AISLE

FINISHED PARTS STORES
1000 SETS BOX PARTS
2000 SETS OF TRAY PARTS

Figure 3-8 Assembly and P.O. Line

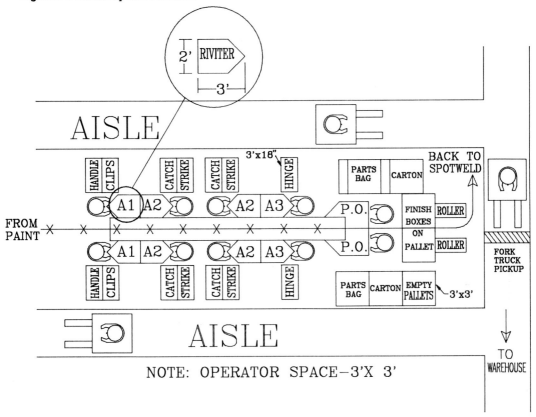

NOTE: OPERATOR SPACE—3'X 3'

Line balancing is an important tool for many aspects of industrial technology, and one of the most important used is the assembly line layout. The back of the assembly line balancing form is designed for an assembly line layout sketch. Look at the examples in Figures 3-7 and 3-8.

Packout work is considered to be the same as assembly work as far as assembly line balancing is concerned. Many other jobs may be performed on or near the assembly line, but are considered sub-assemblies and are not directly balanced to the line, because sub-assemblies can be stockpiled. Their time standards stand on their own merit.

STEP-BY-STEP PROCEDURE FOR COMPLETING THE ASSEMBLY LINE BALANCING FORM

Refer to the assembly line balancing form in Figure 3-9.

① Product No: The product drawing or product part number goes here.

② Date: The complete date of development of this solution.

Figure 3-9 Assembly Line Balancing Step-by-Step Form

FRED MEYERS & ASSOCIATES		ASSEMBLY LINE BALANCING

PRODUCT NO.: ①   PRODUCT DESCRIPTION, ④   "R" VALUE ⑥ EXISTING PRODUCT= $\dfrac{365 \text{ MINUTES}}{\text{UNITS REQ'D/SHIFT}}$ ="R"

DATE: ②

BY I.E.: ③   NUMBER UNITS REQUIRED PER SHIFT ⑤   CALCULATIONS NEW PRODUCT= $\dfrac{300 \text{ MINUTES}}{\text{UNITS REQ'D/SHIFT}}$ ="R"

NO.	OPERATION/DESCRIPTION	"R" VALUE	CYCLE TIME	# STATIONS	AVG. CYCLE TIME	% LOAD	HRS./1000 LINE BALANCE	PCS./HR. LINE BALANCE
⑦	⑧	⑨	⑩	⑪	⑫	⑬	⑭	⑮
							⑯	
			⑲				⑰	

③ By I.E.: The name of the technician doing the assembly line balance—your name.

④ Product Description: The name of the product being assembled.

⑤ Number Units Required Per Shift: This is the quantity of production required per shift—given to the technologist by the sales department. The technologist's objective is to get as close to this quantity as possible without going below.

⑥ R Value: The plant rate has been discussed previously in this chapter, but this block is designed for a specific plant with the following past experience.

a) Existing products have run at 85 percent efficiency.

b) New products average 70 percent efficiency during the first year.

c) Eleven percent allowances are added to each standard. The R value in this plant is calculated by dividing 300 or 365 minutes by the number of units per shift (Step ⑤). The result is the plant rate—the R value.

CHAP. 3: PROCESS DESIGN

(7) No: This is a sequential operation number. The use of operation numbers it to give a simple, useful method of referring to a specific job.

(8) Operation Description: A few well-chosen words can communicate what is being done at this workstation. Parts' names and job functions are the key words. The examples at the back of this chapter are good guides.

(9) R Value: The R value calculated in block (6) above goes behind each operation. The plant rate is the goal of each workstation, and by putting the R value on each line keeps that goal clearly in focus.

(10) Cycle Time: The cycle time is the time standard set by combining elements of work together into jobs. Our goal is the R value, but that specific number can seldom be achieved. Cycle time can be changed by moving an element of work from one job to another, but elements of work are a large proportion of most jobs. Faster equipment or smarter methods may reduce the cycle time, and this is a good cost reduction tool to be talked about later.

(11) # Stations: The number of stations is calculated by dividing the R value (9) into the cycle time (10) and rounding up. If the number of stations is rounded down, the goal (number of units per shift) (5) will not be achieved. Management may round down the number of workstations because of cost, but if they do, they know the goal will not be achieved without overtime, and so on. But that is management's decision, not the technologists. If the number of workstations is rounded down, that workstation will be the bottleneck, the restriction, the slowest station or the 100 percent station.

(12) Ave. Cycle Time: The average cycle time is calculated by dividing the cycle time (10) by the number of workstations (11). This is the speed at which this workstation produces parts. If the cycle time of a job is one minute, and four machines are required, the average cycle time is .250 minutes (1.000 divided by 4 = .250) or a part would come out of those four machines every .250 minutes. The very best line balance would be for every station to have the same average cycle time, but this never happens. A more realistic goal is to work at getting them as close as possible. The average cycle time will be used to determine the percentage workload of each workstation, the next step.

(13) % Load: The percentage load tells how busy each workstation is compared to the busiest workstation. The highest number in the average cycle time column (12) is the busiest workstation and, therefore, is called the *100 percent station*. One hundred percent is written in the percent load column. Now every other station is compared to this 100 percent station by dividing the 100 percent average station time into every other average station time and multiplying the result times by 100 will equal the percent load of each station. The percent load is an indication of where more work is needed or where cost reduction efforts will be most fruitful. If the 100 percent station can be reduced by 1 percent, then we will save 1 percent for every workstation on the line.

> *Example:* To calculate percent load, look back at the example in Figure 3-6 of this chapter. The average cycle times were .153, .146, .130, .119, .155, .139, .125, .138, .125, and .167. Reviewing these average cycle times reveals that .167

is the largest number and is designated the 100 percent workstation. A good practice is to circle the .167 and the 100 percent to remind ourselves that this is the most important workstation on the line, and no other time standard has any further meaning. Now that the 100 percent station is determined, the percentage load of every other workstation is determined by dividing .167 into every other average cycle time:

$$\text{Operation} \quad \text{SSSA1} = .153 \text{ divided by } .167 = 92\%$$

$$\text{SSA1} \quad = .146 \text{ divided by } .167 = 87\%$$

$$\text{SSA2} \quad = .130 \text{ divided by } .167 = 78\%$$

$$\text{Etc.}$$

Where will the supervisor put the fastest person? The P.O. operation! Where will the industrial technologist look for improvement or cost reduction? The P.O. operation, the 100 percent loaded station.

Figure 3-10 Route Sheet—Blank Form

ROUTING SHEET

PART NAME: _____   PART NUMBER: _____

OPERATION #	OPERATION	MACHINE	MACHINE #	MIN. STD.	LINE RATE	MACHINES	PC./HR.	HR./PC.
5								
10								
15								
20								
25								
30								
35								
40								
45								
50								

CHAP. 3: PROCESS DESIGN

A good balance would have all workstations in the 90 to 100 percent range. One workstation below 90 percent can be used for absenteeism. A new person can be put on this station without slowing up the whole line.

(14) Hrs./Unit: The hours per unit produced can most easily be calculated by dividing the 100 percent average cycle time (which is circled on the line balance) by 60 minutes per hour:

$$\frac{.167 \text{ min./unit}}{60 \text{ min./hr}} = .00278 \text{ hrs./unit}$$

The .167 time standard is for one person, so if two people are required on an operation, two times .00278 hours per unit will be required.

$$2 \text{ people} = .00557 \text{ hrs./unit}$$

$$3 \text{ people} = .00834 \text{ hrs./unit}$$

$$4 \text{ people} = .01112 \text{ hrs./unit}$$

Figure 3-11 Assembly Line Balancing—Blank Form

	LINE BALANCING		DATE:			PAGE	OF	PAGES

	# PARTS REQ/DAY–SHIFT	% PLANT PERFORMANCE		DOWN TIME	"R" VALUE	HRS./1,000/PERSON	
NO.	OPERATION/DESCRIPTION	"R" VALUE	CYCLE TIME	# STATIONS	AVG. CYCLE TIME	% LOAD	HRS. PER 1,000	PIECES/ HOUR	HRS./1000 LINE BALANCE	PCS./HR. LINE BALANCE

Also, as a quality check, if you multiply the number of operators on your assembly line times .00278 hours per unit you get the total hours to make one unit. In our tool box problem (25 × .00278 = .06950) hours of labor will be required to assemble each tool box. Another piece of logic is that everyone on an assembly line must work at the same rate. The person with the least work to do still cannot do more than comes to the operator, and cannot do one more than the following operator can do.

⑮ Pieces/Hr.: Pieces per hour is $1/X$ of hrs./unit, or divided the hours per unit into one. Notice on the example in Figure 3-6, all the stations produce 360 pieces. Station A1 has two operators, each producing 180 pieces per hour for 360 pieces per hour total.

⑯ Total Hrs./Unit: Total hours per unit is the number of hours adding all the operations together. The hours per unit for one operator times the total number of operators on the line also equals the total hours per unit. The total of column ⑪ is the total operators.

Figure 3-12 Fabrication Equipment Spreadsheet—Blank Form

FABRICATION EQUIPMENT SPREADSHEET

MACHINES																		MACH. REQ'D	TOT. # MACH.
BREAK PRESS–FORM																			
BREAK PRESS–DRAW																			
BREAK PRESS–EDGES																			
VAPOR DEGREASER SOLVENT SPRAY																			
SHOT BLAST MACH.																			
ALKALINE SOAP/ WATER SPRAY																			
DRYERS																			
TUBE BENDER																			
BEND MACHINE																			
HELI ARCH																			
INJECTION MOLD																			
DOUBLE DRILL PRESS																			
ABRASION SAW																			
PAINT BOOTHS																			
OVENS																			

CHAP. 3: PROCESS DESIGN

⑰ Average Hourly Wage Rate: This would come from the payroll department, but for example, let's say $7.50/hr. is the average hourly wage rate.

⑱ Labor Cost Per Units: In our example, .06950 hours times $7.50/hr. = $.52 per units labor cost. The lower the cost, the better line balance.

⑲ Total Cycle Time: The total cycle time tells us the exact work content of the whole assembly, and if treated like any other time standard, can show us what a perfect line balance would be.

Our example 3.494 minutes divided by 60 mins./hr. equals .05823 hours per unit. Our line balance came out .06950, or .01127 hours more. This .01127 hours is potential cost reduction, and what cannot be removed by cost reduction is called the cost of line balance.

LAYOUT ORIENTATION

Mass production and job shop are the two basic layout orientations. *Mass production* is product-oriented and follows a fixed path through the plant. The assembly line best illustrates the mass production orientation. Mass production orientation is preferred over job shop because the unit cost is lower, but not all product follows a fixed path.

The *job shop* orientation layout is process-oriented (built around machine centers). Fabrication departments are usually layed out this way because the paths taken by the parts are not consistent. This is called a *variable path flow*.

Because mass production is preferred over job shop, several new techniques have been developed to move job shop orientation closer to mass production:

1. Group technology tries to classify parts into groups with similar process sequences. The equipment then can be placed in a straight line approaching a fixed path. Our plant could have a sheet metal line, plastics line, bar stock line, casting line, and so on. Reducing cross traffic, backtracking and feet of travel is our objective.

2. A work cell is a group of machines to make one complicated part. One or two operators may run six to ten machines. The machines are dedicated to this one part and remained set up for this part forever. Some machines may not be fully utilized, but the lost time on these machines is counterbalanced by less inventory required, less material handling, and much shorted through put time (the time a part spends in production).

Most plants use job shop orientation for the fabrication end of the plant and mass production orientation for assembly lines and packout. As opportunities arise group technology and work cells are created.

In the next chapter, flow analysis techniques will be discussed to optimize layouts of fabrication and assembly areas.

QUESTIONS

1. What is process design?
2. What are the two categories of process design?
3. What is a route sheet?
4. What information is included on a route sheet?
5. What determines how many machines we buy?
6. Which time standard (decimal minute, pieces per hour, or hours per unit) compares to the R value?
7. What is an assembly chart (see Figure 3-4)?
8. What information is needed to calculate an assembly line conveyor speed?
9. What additional information is needed to calculate paint conveyor speed?
10. What are the eight purposes of assembly line balancing?
11. Rebalance Figure 3-6 by adding a fourth packout person and make a sub-assembly out of SA3 and answer the following.
    a) The total hrs./unit?_____
    b) How many units per shift will be made @ 100 percent?_____
    c) How many people are now used?_____
    d) What is the new 100 percent station?_____
    e) Is this a better balance?_____
    f) How much money is saved if we produce 700,000 units per year and the employees are paid $10.00/hour?_____
12. What are the two primary layout orientations?
13. Balance the following assembly line to produce 1,500 units per shift @ 85 percent and 30 minutes personal time.

Operation #	Time Standard
1	.390
2	.235
3	.700
4	1.000
5	.240
6	.490

Chapter 4

Flow Analysis Techniques

Flow analysis is the guts of plant layout and the beginning of the material handling plan. The *flow* of a part is the path which that part takes while moving through the plant. *Flow analysis* considers the path that every part takes through the plant and tries to minimize the:

1. distance travelled (measured in feet);
2. backtracking;
3. cross traffic; and
4. cost of production.

In a positive vein, flow analysis will assist in the most effective arrangement of machines, workstations, and departments. It is said that if you improve the product flow, you will automatically increase profitability. We can improve flow by developing product or parts classes with similar flow. We can try to get each part to take a similar path, and we can try to move the parts automatically.

Our employees have flow as well, so we want to consider their paths. For example, most employees drive to work, park their cars, enter through the employee entrance, punch their time cards, go to their lockers, go to the cafeteria, and then go to their workstation. We want to use this flow to place these services conveniently.

Two basic groups of flow analysis techniques are individual parts flow through fabrication and total plant flow.

The study of individual parts flow results in the arrangement of machines and workstations. Route sheets are our primary source of information. To establish this best arrangement of equipment, we will use four techniques:

1. A string diagram;
2. A multi-column process charts;
3. A from-to chart; and
4. A process chart.

We may not use each technique every time, but using more than one technique is a good practice. To show how these techniques work, let's consider a small group of parts (see Figure 4-1) with the following routing (flow). This flow routing will be considered inflexible so that we must layout (or relayout) the workstations. We need 2,000 units per day of all parts and the parts weigh .5 lbs., 9 lbs., 5 lbs., 15 lbs., and 3.75 lbs., respectively. The machines are identified with a letter (R A B C D E F S). R is the incoming material location (called receiving), and S is the shipping end of the line. Using my most creative ability to layout these machines, I'll put them in alphabetical order first, then check the efficiency.

String Diagram

We will let circles represent our equipment and the line between circles will indicate flow (see Figures 4-2 and 4-3). Flow lines between adjacent circles will be from middle of circle to middle of circle. If we jump a department we will place the line above the circles. If the flow is backward, called *backtracking* (going toward R), the flow line is drawn under the circles (see Figure 4-2).

Look at the important relationships (the two circles with a lot of lines between them). What is clear in the string diagram above is that this arrangement of machines produces a lot of travel. To improve this layout, look for important relationships (more than one part taking the same route). Look at the C to D relationship. Four of the five parts make this trip, so this relationship is important and C must stay close to D. Here are other important relationships.

Figure 4-1 Routing for Five Parts

Part #	Routing (Operation Sequence)
1	R A B D C F S
2	R B D C A S
3	R E F B A C D S
4	R F A C D S
5	R C A D S

Figure 4-2 String Diagram (Alphabetical Layout—First Try)

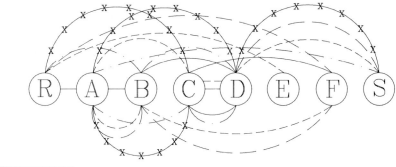

1. ──────
2. — ·— ·— ·—
3. ------------
4. — — — —
5. ×─×─×─×──

1. B–D has 2 lines.
2. A–C has 4 lines.
3. D–S has 3 lines.

So let's rearrange the layout (see Figure 4-3). Which is best? We could move each part seven steps from R to S, so a perfect layout would require moving only 7 steps × 5 parts = 35 steps. A step is the distance between the center of one circle

Figure 4-3 String Diagram—Improved Method

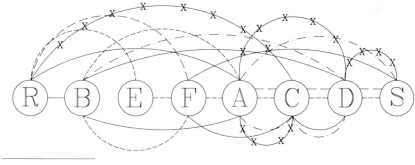

1. ──────
2. — ·— ·— ·—
3. ------------
4. — — — —
5. ×─×─×─×──

to the center of the adjacent circle. If we jump one circle, two steps would be required.

In our first alphabetical layout, part 1 went from R to A to B to D to C to F to S for a total of 9 steps. Part 2 traveled 13 steps; part 3 traveled 17 steps; part 4 traveled 17 steps; and part 5 traveled 11 steps.

Part #	# Steps Traveled
1	9
2	13
3	17
4	17
5	11
TOTAL	67

$$\text{Efficiency} = \frac{35}{67} = 52\%$$

The second layout produced less steps:

Part #	# Steps Traveled
1	19
2	11
3	11
4	7
5	9
TOTAL	57

$$\text{Efficiency} = \frac{35}{57} = 61\%$$

How efficient can you make this layout?

Multi-Column Process Chart

Using the same routing information used in the string diagram for our 5 parts, we can show the flow for each part right next to but separate from each other (see Figures 4-4 and 4-5). First of all, we list the operations down the left side of the page, then set up a column next to the list of operations—one for each part as follows (see Figure 4-4).

$$\text{Efficiency} = \frac{35}{67} = 52\%$$

Figure 4-4 Multi-Column Process Chart

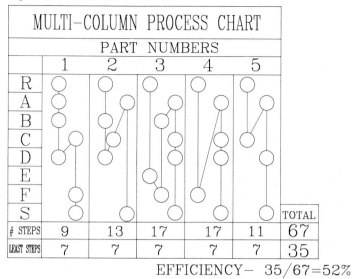

MULTI-COLUMN PROCESS CHART

	PART NUMBERS					
	1	2	3	4	5	
R						
A						
B						
C						
D						
E						
F						
S						TOTAL
# STEPS	9	13	17	17	11	67
LEAST STEPS	7	7	7	7	7	35

EFFICIENCY— 35/67=52%

This is the same efficiency we established in the string diagram. Let's try to improve it again, but different (better) than our second string diagram. Look for clues to improvements. Figure 4-5 is an improved layout.

$$\text{Efficiency} = \frac{35}{49} = 71\%$$

Figure 4-5 Multi-Column Process Chart—Improved Layout

MULTI-COLUMN PROCESS CHART

	PART NUMBERS					
	1	2	3	4	5	
R						
E						
F						
B						
A						
C						
D						
S						TOTAL
# STEPS	17	11	7	7	9	49
LEAST STEPS	7	7	7	7	7	35

EFFICIENCY—35/49=71%

FABRICATION OF INDIVIDUAL PARTS

We have come a long way toward a perfect layout. My alphabetical layout produced only a 52 percent efficiency, now we are 71 percent efficient. We can still get better. How good can you do? Before you try another improvement, study the third technique of flow analysis, the from-to chart.

From-To Chart

The from-to chart is the most exact technique of the three. We can develop an efficiency that considers the importance of the parts. Up until now, we have considered each part as equal in importance, but at the beginning of this chapter the quantity and weight of each part was recorded. Figure 4-6 is a chart of this data given earlier.

Routing for Five Parts*

Part #	Routing (Operation Sequence)
1	R A B D C F S
2	R B D C A S
3	R E F B A C D S
4	R F A C D S
5	R C A D S

* From Figure 4-1.

The relative importance of part 4 is 30 times more important than part 1 and 3, so it should have 30 times more effect on our layout.

The from-to chart is a matrix. The sequence of operations is written down the left side of the form and across the top. The vertical sequence of machines is the "from" side of our matrix. The horizontal sequence of machines is the "to" matrix. Everything moves *from* some place *to* some place. Each time a move is required, a weighted value is placed in that coordinate (see Figure 4-7). For an example of all five parts, see Figure 4-8.

To evaluate this alternative, penalty points are assigned to each move depending upon how far the move is away from the present location. For example, the move R to A is right next door, so we multiply that weight time 1 (one block). R to B is two blocks away, so we multiply the 18 in that block times two, three blocks away times three, and so on. In Figure 4-9, the circled numbers are the

Figure 4-6 Part Quantity and Weight Data

Part #	Quantity Per Day	Weight in Pounds	Weight	Relative Importance*
1	2,000	0.5	1,000#	1.0
2	2,000	9.0	18,000#	18.0
3	2,000	0.5	1,000#	1.0
4	2,000	15.0	30,000#	30.0
5	2,000	3.75	7,500#	7.5

* These numbers and the routing in Figure 4-1 creates the weighted value of each move.

Figure 4-7 From-To Chart—Example of Part 1 With a Relative Value of 1

T O

	R	A	B	C	D	E	F	S
R		1						
A			1					
B					1			
C							1	
D				1				
E								
F								1
S								

(Left margin label: **F R O M**)

Figure 4-8 From-To Chart—Alphabetical Layout

T O

	R	A	B	C	D	E	F	S	Total
R		1	18	7.5		1	30		57.5
A			1	1+30	7.5			18	57.5
B		1			1+18				20
C		18+7.5			30+1			1	57.5
D				18+1				30+1 7.5	57.5
E							1		1
F		30	1					1	32
S									

(Left margin label: **F R O M**)

TOTAL: 283

Figure 4-9 From-To Chart—Alphabetical Layout Analysis

T O

		R	A	B	C	D	E	F	S	T	P.P.
	R		1 (1)	18 (36)	7.5 (22.5)		1 (5)	30 (180)		57.5	244.5
	A			1 (1)	31 (62)	7.5 (22.5)			18 (108)	57.5	193.5
F	B		1 (2)			19 (38)				20	40
R	C		25.5 (102)			31 (31)			1 (3)	57.5	136
O	D				19 (38)				38.5 (115.5)	57.5	153.5
M	E							1 / 1		1	1
	F		30 (300)	1 (8)					1 (1)	32	309
	S									—	—
	T P.P.		57.5	20	57.5	57.5	1	32	57.5	283	1077.5

P.P. = Penalty Points

penalty points. Below and to the left of the diagonal line indicates backtracking, so the penalty points are doubled.

The efficiency of the alphabetical layout is:

$$\frac{283}{1,077.5} = 26\%$$

Now look for clues to improvement. The highest penalty points are the best clues. For example, from F to A has a penalty point of 300. This means that F wants to be closer to A. The move of R and F has a penalty point of 180. This means F wants to be closer to R. A new layout change of sequence will change both the vertical and horizontal sequence. Figure 4-10 is an improved layout.

$$\text{Efficiency } \frac{283}{548} = 51.6\%$$

Figure 4-10 From-To Chart—Alphabetical Layout Analysis

T O

FROM \ TO	R	E	F	B	A	C	D	S	T	P.P.
R		1 ⓵	30 ㊿(60)	18 (54)	1 (4)	7.5 (37.5)			57.5	156.5
E			1 (1)						1	1
F				1 (1)	30 (60)			1 (5)	32	66
B					1 (1)	1+18 (57)			20	58
A				1 (2)		1+30 (31)	7.5 (15)	18 (54)	57.5	102
C			1 (6)		18+7.5 (51)		1+30 (31)		57.5	88
D						1+18 (38)		1+30+7.5 (38.5)	57.5	76.5
S									—	—
T		1	32	20	57.5	57.5	57.5	57.5	283	
P.P.		1	67	57	116	106.5	103	97.5		548.0

This still can be improved. Find the best layout. A 56 percent is the best possible. A perfect layout (straight through flow—no backtracking) is not possible in this case because of the different routing for each part. We are assuming there is no other routing possible. If we could change the routing of even one part, we could improve the efficiency. Practical limitations may dictate the routing, so we are left with the need to arrange the machines and equipment the best possible way. The string diagram, multicolumn process chart and the from-to chart are techniques to help us find that best layout.

The first part of this chapter listed four goals of flow analysis. Minimizing distance traveled and backtracking are the first two goals. These first three techniques (string diagram, multi-column process chart, and from-to chart) address these goals. Discouraging backtracking was best addressed by these techniques, but distances were considered only in relative terms. Future techniques will allow us to calculate exact distance measured in feet. Minimizing the cost of production is the ultimate goal of flow analysis. The final fabrication flow analysis technique speaks to this point.

Process Chart

The process chart (see Figure 4-11) is used for just one part, recording everything that happens to that part from the time it arrives in the plant until it joins the other parts. Symbols are used to describe what happens:

Symbols		Description
○	=	Operation, Work on the Part
⇨	=	Transportation, Moving the Part
▽	=	Storage, Storerooms, Warehouse, Work in Progress
D	=	Delay, Very Temporary Storage, Usually at a Workstation Both Incoming Containers and Outgoing Containers of Parts
□	=	Inspection, Q.C. Work on the Product
⬭	=	Combination Operation and Inspection

Process charting lends itself to a standard form. A properly designed form will lead the designer to ask questions of each step. The designer wants to know the why, who, what, where, when and how of every operation, transportation, inspection, storage, and delay. Once the designer understands the answer to these questions, he/she can ask:

1. Can I eliminate this step?
2. Can I automate this step?
3. Can I combine this step with another?
4. Can I change the routing to reduce distances traveled?
5. Can I move workstations closer together?
6. Can I justify production aids to increase effectiveness?
7. How much does this part cost to produce?

Step-by-Step Description for the Process Chart

This step-by-step procedure accompanies Figure 4-11.

① [   ] Present Method (or) [   ] Proposed Method

Figure 4-11 Sample Process Chart

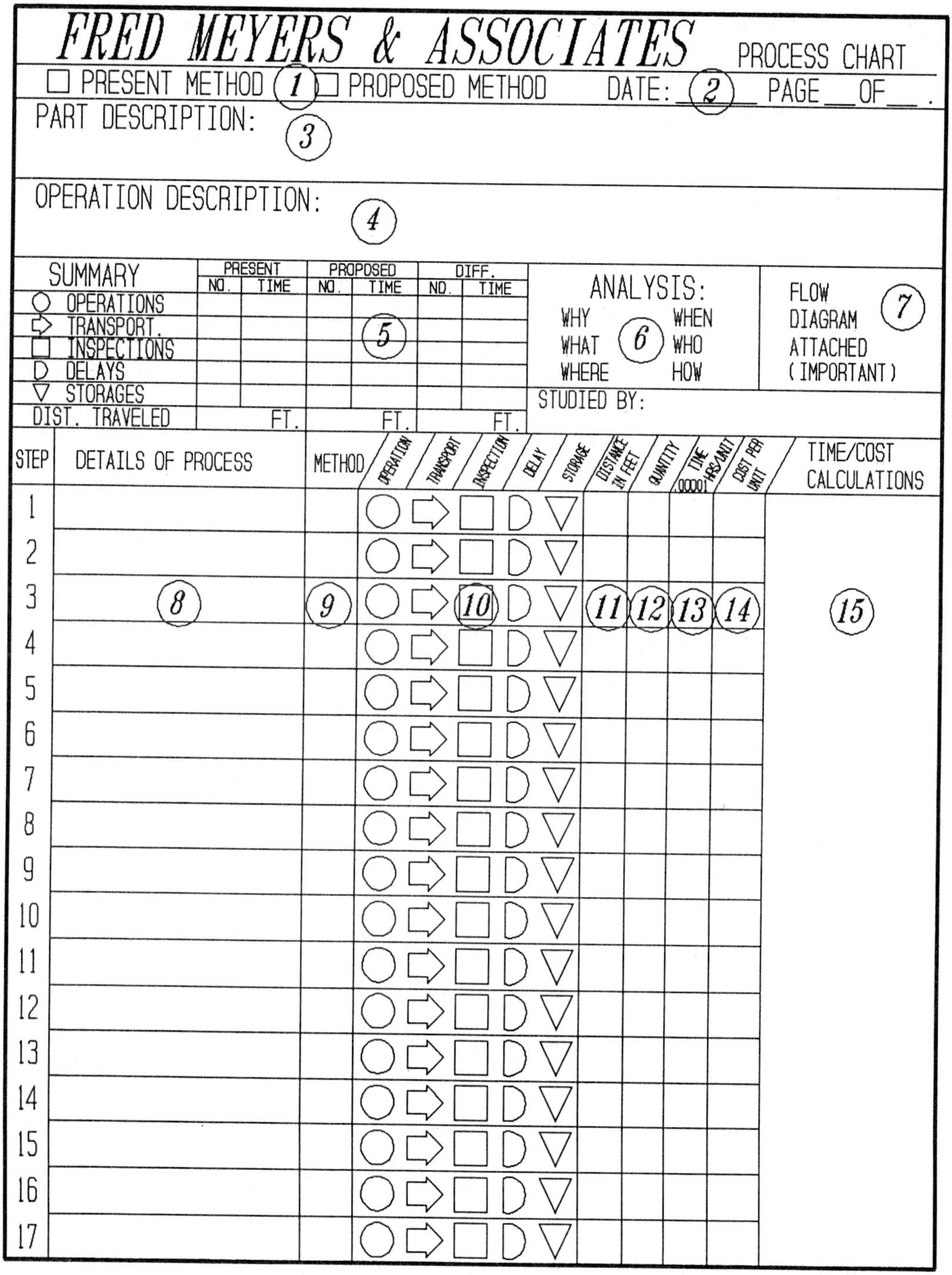

FRED MEYERS & ASSOCIATES

PROCESS CHART

□ PRESENT METHOD (1) □ PROPOSED METHOD   DATE: (2)   PAGE ___ OF ___ .

PART DESCRIPTION: (3)

OPERATION DESCRIPTION: (4)

SUMMARY	PRESENT		PROPOSED		DIFF.	
	NO.	TIME	NO.	TIME	NO.	TIME
○ OPERATIONS						
⇨ TRANSPORT.			(5)			
□ INSPECTIONS						
D DELAYS						
▽ STORAGES						
DIST. TRAVELED		FT.		FT.		FT.

ANALYSIS:
WHY   WHEN
WHAT (6) WHO
WHERE   HOW
STUDIED BY:

FLOW
DIAGRAM (7)
ATTACHED
(IMPORTANT)

STEP	DETAILS OF PROCESS	METHOD	OPERATION	TRANSPORT	INSPECTION	DELAY	STORAGE	DISTANCE IN FEET	QUANTITY	TIME .00001	COST PER UNIT	TIME/COST CALCULATIONS
1			○	⇨	□	D	▽					
2			○	⇨	□	D	▽					
3	(8)	(9)	○	⇨	(10)	D	▽	(11)	(12)	(13)	(14)	(15)
4			○	⇨	□	D	▽					
5			○	⇨	□	D	▽					
6			○	⇨	□	D	▽					
7			○	⇨	□	D	▽					
8			○	⇨	□	D	▽					
9			○	⇨	□	D	▽					
10			○	⇨	□	D	▽					
11			○	⇨	□	D	▽					
12			○	⇨	□	D	▽					
13			○	⇨	□	D	▽					
14			○	⇨	□	D	▽					
15			○	⇨	□	D	▽					
16			○	⇨	□	D	▽					
17			○	⇨	□	D	▽					

A checkmark in one of the two boxes is required. A good industrial technologist's practice is to always record the present method so that the improved (proposed) method can be compared. Costing the present and proposed methods will be required to justify your proposal, especially if any costs are involved. Recording and advertising cost reduction dollars saved is a smart idea for any technologist.

② Date _____ Page _____ of _____: Always date your work. Our work tends to stay around for years, and you will someday want to know when you did this great work. Page numbers are important on big jobs to keep the proper order.

③ Part Description: This is probably the most important information on the form. Everything else would be useless if we didn't record the part number. Each process chart is for one part, so be specific. The part description also includes the name and specifications of the part. Attaching a blueprint to the process chart would be useful.

④ Operation Description: In this block, you record the limits of the study, for example, from Receiving to Assembly. Also, any miscellaneous information can be placed here.

⑤ Summary: The summary is used only for the proposed solution. A count of the operations, transportation, inspection, delays, and storage for the present and proposed methods are recorded and the difference (savings) is calculated.

The distance traveled is calculated for both methods and the difference calculated. The time standards in minutes or hours is summarized and the difference calculated. This information is why we did all the work of present and proposed process charting; it is the cost reduction information. We will come back to Step 5 after Step 15.

⑥ Analysis: "These are the six honest serving men; they've taught me all I know. Their names are why, what, where, when, how, and who." These questions are asked of each step (line) in the process chart. "Why" is first. If we don't have a good why, we can eliminate that step and save 100 percent of the cost. The questioning of each step is how we come up with the proposed method. With these questions, we try to:

a) Eliminate every step possible, because this produces the greatest savings.

b) But, if we can't eliminate the step, we try to combine steps to spread the cost and possibly eliminate steps between. For example, if two operations are combined, delays and transportation can be eliminated. If transportation are combined, many parts will be handled as one.

c) But, if we can't eliminate or combine, maybe we can change the sequence of operations around to improve and reduce flow and save many feet of travel.

As you can see, the analysis phase of process charting gives the process meaning and purpose. We should come back to Step 6 after Step 15 for cost reduction.

⑦ Flow Diagram Attached (IMPORTANT): Process charting is used in conjunction with flow diagramming. The same symbols can be used in both techniques. The process chart is the words and numbers, where the flow diagram is the picture. The flow diagram is the next technique. The present and proposed methods of both techniques must be telling the same story; they must agree.

Studied by: is where *your* name goes. Be proud and put your signature here.

⑧ Details of Process: Each line in the flow process chart is numbered, front and back. One chart can be used for 42 steps. Each step is totally independent and stands alone. A description of what happens in each step aids the analyst's questions. Using as few words as possible, describe what is happening. This column is never left blank.

⑨ Method: Method usually refers to how the material was transported—by fork truck, hand cart, conveyer, hand—but methods of storage could also be placed here.

⑩ Symbols: The process chart symbols are all here. The analyst should classify each step and shade the proper symbol to indicate to everyone what this step is.

⑪ Distance in Feet: This step is used only with the transportation symbol. The sum of this column is the distance traveled in this method. This column is one of the best indications of productivity.

⑫ Quantity: Quantity refers to many things:

a) Operations: The pieces per hour would be recorded here.

b) Transportation: How many were moved at a time.

c) Inspection: How many pieces per hour if under time standard and/or frequency of inspection.

d) Delays: How many pieces in a container. This will tell us how long the delay is.

e) Storage: How many pieces per storage unit.

All costs will be reduced to a unit cost or cost per unit, so knowing how many pieces are moved at one time is important.

⑬ Time in Hours Per Unit (.00001): This step is for labor costing. Storage and delays will be costed in another way—inventory carrying cost. This column will be used only for operations, transportation, and inspection. Time per unit is calculated in two ways:

a) Starting with pieces per hour time standards, say 250 pieces per hour, divide 250 pieces per hour into 1 hour, and you get .00400 hours per unit. On our process chart, we place 400 in the time column, knowing that the decimal is always in the fifth place.

b) Starting with a material handling time of 1.000 minutes to change a tub of parts at a workstation with a hand truck, and we have 200 parts in that tub. How many hours per unit is our time standard?

$$\frac{1.000 \text{ min./container}}{200 \text{ parts/container}} = .005 \text{ min./part}$$

$$\frac{.005 \text{ min./part}}{60 \text{ min./hr.}} = .00008 \text{ hr./part}$$

⑭ Cost Per Unit: Hours per unit multiplied by a labor rate per hour equals a cost per unit. For example, the above two problems using a labor rate of $7.50 per hour:

a) .00400 × 7.50 = $ .03 per unit

b) .00008 × 7.50 = $ .0006 per unit

The cost per unit is the backbone of our process charting. We are looking for a better way, so the method with the overall cheapest way is the best method.

⑮ Time/Cost Calculations: Technologists are required to calculate costs on many different things, and how costs were calculated tends to get lost. This space is provided to record the formulas developed to determine the costs so that they do not have to be redeveloped over and over again.

⑤ Returning Summary: Once all steps in the present method process chart have been completed, the summary is completed by:

a) counting all the operations, transportation, etc.

b) adding up the unit time for all steps

c) adding up the distance traveled.

TOTAL PLANT FLOW

The three techniques studied in this section are:

1. flow diagram;
2. operations chart; and
3. flow process chart.

We will consider every step in the process for fabrication, assembly, and packout of the product. The techniques use the same symbols as used in the process chart, but in a different way. All parts are considered, not just one.

Flow Diagrams

The *flow diagrams* (see Figure 4-12 and 4-13) show the path traveled by each part from receiving to stores to fabrication of each part to subassembly to final assembly to packout to warehousing to shipping. These paths are drawn on a layout of the plant.

The flow diagram will point out problems with such things as cross traffic, backtracking, and distance traveled.

Figure 4-12 Flow Diagram

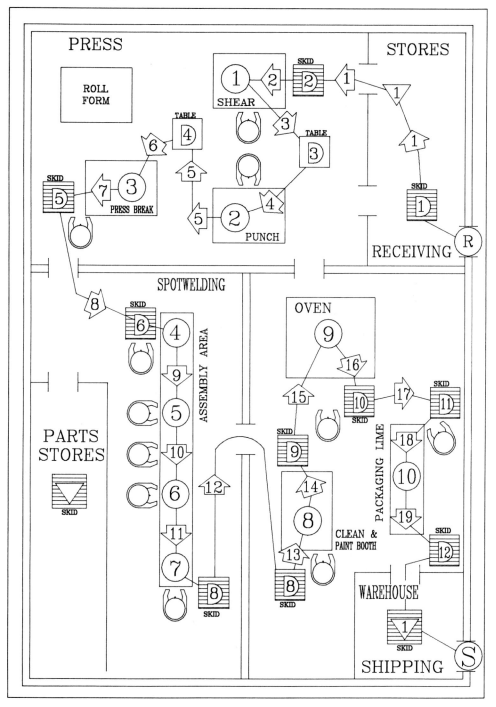

Figure 4-13 Flow Diagram—Tool Box Plant

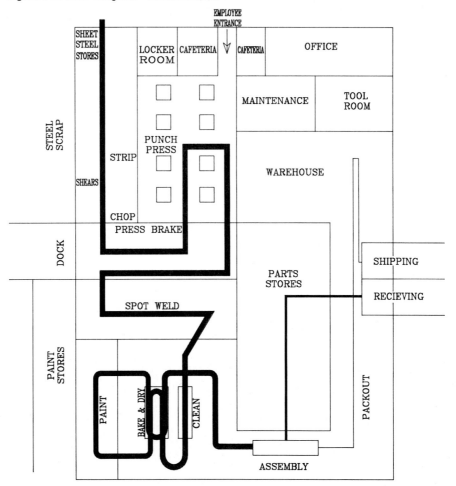

Cross traffic. Cross traffic is where flow lines cross. Cross traffic is undesirable and a better layout would have less intersecting paths. Anywhere traffic crosses is a problem because of congestion and safety considerations. Proper placement of equipment, services, and departments will eliminate most cross traffic.

Backtracking. Backtracking is material moving backwards in the plant. Material should always move toward the shipping end of the plant. If it is moving toward receiving, it's moving backwards. Backtracking costs three times as much as flowing correctly. For example, five departments have a flow like this:

How many times did material move between departments 3 and 4? Three times! Twice forward and once backward. If we arranged this plant and changed departments 3 and 4 around, we would have straight through flow:

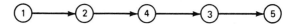

This arrangement has no backtracking! Efficiency wise, we travel less distance. In our first example, we traveled six blocks (a block is one step between departments next to each other). In the straight line flow, we traveled only four blocks—a 33 percent increase in productivity.

Distance traveled. Distance costs money to travel. The less distance traveled, the better. The flow diagram is developed on a layout, and the layout can be easily scaled and the distance of travel calculated. By rearranging machines or departments, we may be able to reduce the distances traveled.

Because flow diagrams are created on plant layouts, no standard form is used. There are few conventions to restrict the designer. The objective is to show all the distances traveled by each part and to find ways of reducing the overall distance.

The flow diagram is developed from routing sheet information, assembly line balance, and blueprints. The routing sheet specifies the fabrication sequence for each part of a product. This sequence of steps required to make a part is very practical and has some room for flexibility. One step may come before or after another step, depending on conditions. The sequence of steps should be changed to meet the layout if impossible, because that requires only a paperwork change. But if the sequence of operations cannot be changed and the flow diagram shows backtracking, moving equipment may be necessary. Our objective will always be "make a quality part the cheapest, most efficient way possible."

Step-by-Step Procedure for Developing a Flow Diagram

Step 1: The flow diagram starts with an existing or proposed scaled layout.

Step 2: From the route sheet, each step in the fabrication of each part is plotted and connected with a line, use color codes, or other method of distinguishing between parts.

Step 3: Once all the parts are fabricated, they will meet, in a specific sequence, at the assembly line. The position of the assembly line will be determined by where the individual parts came from. At the assembly line, all flow lines join together and travel as one to packout, warehouse, and shipping. A well thought-out flow diagram will be the best technique for developing a plant layout.

Plastic overlays on plant layouts are often used to develop flow lines for flow diagrams. The flow lines can be drawn with a grease pencil and can be grouped by classes for plants with a lot of different parts. It doesn't take a large product to make the fabrication departments of a plant layout look like a bowl of spaghetti. Using several plastic overlays will simplify the analysis.

A new industrial technologist will learn much from the creation of a flow diagram, and an experienced technologist will always find ways to improve the flow of material.

Figure 4-12 shows the flow of one part through a plant. Can you recommend improvements? Figure 4-13 shows the flow in our tool box example.

The Operations Chart

The *operations chart* (see Figures 4-14a and 4-14b) has a circle for each operation required to fabricate each part, to assemble each part to the final assembly, and to packout the finished product. On one piece of paper, every production operation, every job, and every part is included.

Operations charts show the introduction of raw materials at the top of the page, on a horizontal line. (See Figure 4-14a.) The number of parts will determine the size and complexity of the operations chart.

Below the raw material line, a vertical line will be drawn connecting the circles (steps in the fabrication of that raw material into finishes parts). Figure 4-14a illustrates these points. Once the fabrication steps of each part is plotted, the parts flow together in assembly. Usually, the first part to start the assembly is shown at the far right of the page. The second part is shown to the left of that, and so forth working from right to left (see Figure 4-14b). Some parts require no fabrication steps. These parts are called *buyouts*. Buyout parts are introduced above the operation at which they will be used (shown on the bottom of Figure 4-14b—Packout Operation). In the packout operation, we are going to place six products into a master carton and tape it closed.

The operations chart shows a lot of information on one page. The raw material, the buyouts, the fabrication sequence, the assembly sequence, the equipment needs, the time standards, and even a glimpse of the plant layout, labor costs, and plant schedule can all be derived from the operations chart. Is it any wonder that plant layout designers consider this one of their favorite tools?

The operations chart is different for every product, so a standard form is not practical. The circle is universally accepted as the symbol for operations, thereby the origin of the chart's name. There is more convention in operations charting than in flow diagramming, but the designers should not be too rigid in their thinking.

Step-by-Step Procedures for Preparing an Operations Chart

Step 1: Identify the parts that are going to be manufactured and those that are going to be purchased complete.

Step 2: Determine the operations required to fabricate each part and the sequence of these operations.

Figure 4-14a Sample Operation Chart

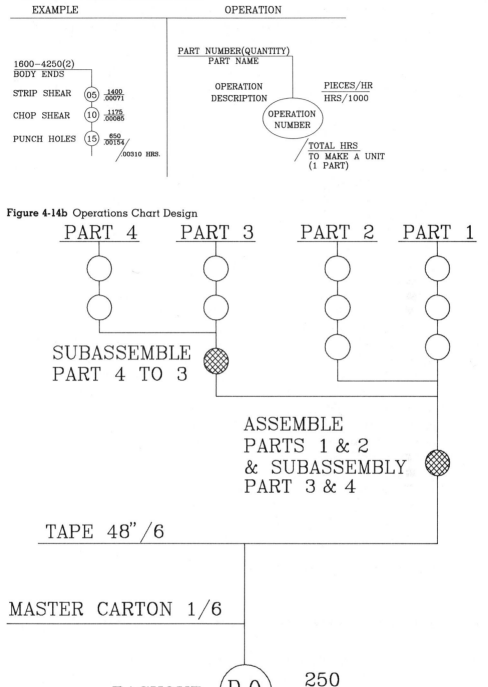

Figure 4-14b Operations Chart Design

Figure 4-14c Sample Operations Chart

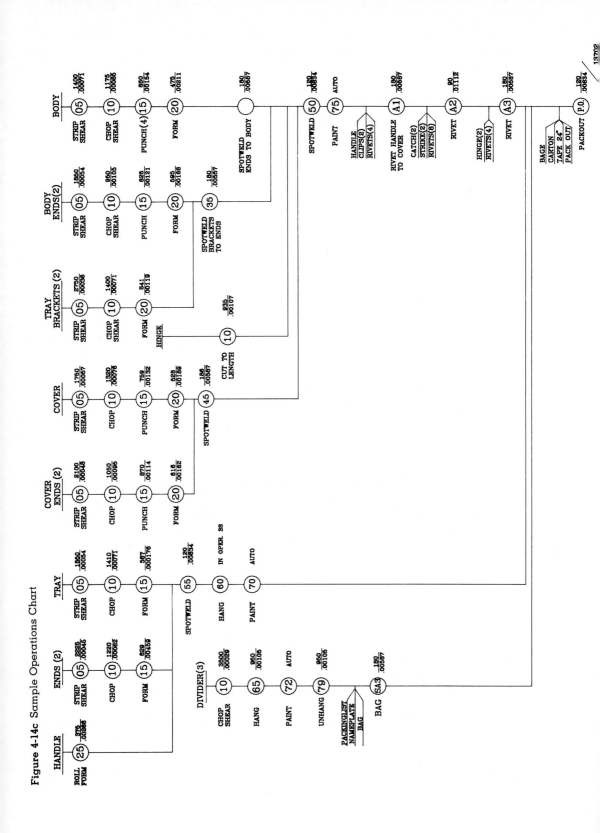

Step 3: Determine the sequence of assembly, both buyout and fabricated parts.

Step 4: Find the base part. That is the first part that starts the assembly process. Put that part on a horizontal line at the far right side top of your page. On a vertical line extending down from the right side of the horizontal line, place a circle for each operation. Beginning with the first operation, list all operations down to the last operation.

Step 5: Place the second part to the left of the first part and the third part to the left of the second part, and so forth until all manufactured parts are listed across the top of the page in reverse order of assembly. All of the fabrication steps are listed below the parts with a circle representing each operation.

Step 6: Draw a horizontal line from the bottom of the last operation of the second part to the first part just below its final fabrication operation and just above the first assembly operation. Depending on how many parts the first assembler puts together, the third, fourth, etc., parts will flow into the first parts vertical line, but always above the assembly circle for that assembly operation.

Step 7: Introduce all buyout parts on horizontal lines above the assembly operation circle where they are placed on the assembly.

Step 8: Put time standards, operation numbers, and operation descriptions next to and in the circle as explained earlier.

Step 9: Sum total the hours per unit and place these total hours at the bottom right under the last assembly or packout operation.

Figure 4-14c is a good example of an operations chart showing *sub-assembly*. Some parts will flow together before they reach the assembly line. This could be welding parts together or assembling a bag of parts. This is called *sub-assembly* and is treated just like the main assembly except it is done before the parts reach the far right side vertical line. Bag packing is a good example. All parts are usually buyouts and could be placed at the bottom left of your operations chart like bag packing in Figure 3-4 SA3 and Figure 4-14 SA3.

Flow Process Chart

The *flow process chart* combines the operations chart with the process chart. The operations chart used only one symbol—the circle or operations symbol. The flow process chart is just five times more, using all five process chart symbols. Another difference is that buyout parts are treated like manufactured parts. No standard form exists to flow process charting (see Figure 4-15).

The flow process chart is the most complete of all the techniques, and when completed, the technologist will know more about the plant's operation than anyone in the plant.

Figure 4-15 Flow Process Chart

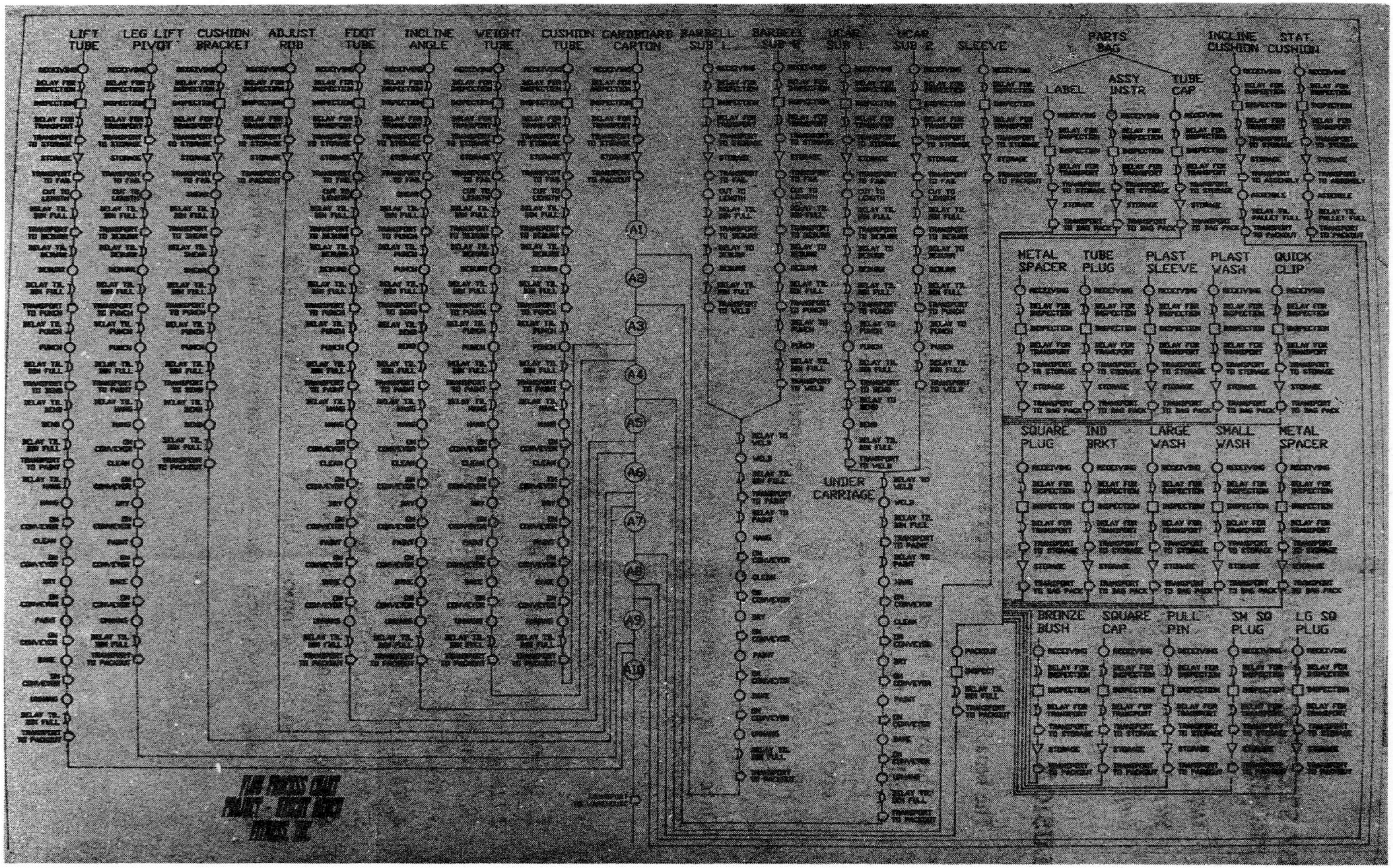

Step-by-Step Procedure for Preparing a Flow Process Chart

Step 1: Start with an operations chart.

Step 2: Complete a process chart for each part.

Step 3: Combine the operations chart and the process chart working in all the buyouts.

CONCLUSION

Flow analysis leads to better plant layouts. Promoting efficiency, effectiveness, and cost reduction are the goals of flow analysis. A detailed analysis of material flow will arm the layout designer with critical information such as

1. operation requirements;
2. material handling needs;
3. storage needs;
4. inspection requirements; and
5. delay reasons.

With this information, the designer is challenged to:

1. Eliminate as many steps as possible.
2. Combine steps.
3. Rearrange equipment to:
   a) eliminate cross traffic;
   b) eliminate backtracking; and
   c) reduce the distance of travel.
4. In general, reduce production costs.

Figures 4-16, 4-17, and 4-18 are blank forms for your use.

Figure 4-16 Process Chart

FRED MEYERS & ASSOCIATES
PROCESS CHART

☐ PRESENT METHOD   ☐ PROPOSED METHOD   DATE:_____ PAGE__OF__ .

PART DESCRIPTION:

OPERATION DESCRIPTION:

SUMMARY	PRESENT NO.	PRESENT TIME	PROPOSED NO.	PROPOSED TIME	DIFF. NO.	DIFF. TIME	ANALYSIS:		FLOW
◯ OPERATIONS							WHY	WHEN	DIAGRAM
⇨ TRANSPORT.							WHAT	WHO	ATTACHED
☐ INSPECTIONS							WHERE	HOW	(IMPORTANT)
D DELAYS							STUDIED BY:		
▽ STORAGES									
DIST. TRAVELED		FT.		FT.		FT.			

STEP	DETAILS OF PROCESS	METHOD	OPERATION	TRANSPORT	INSPECTION	DELAY	STORAGE	DISTANCE IN FEET	QUANTITY	TIME HRS/UNIT /0000I	COST PER UNIT	TIME/COST CALCULATIONS
1			◯	⇨	☐	D	▽					
2			◯	⇨	☐	D	▽					
3			◯	⇨	☐	D	▽					
4			◯	⇨	☐	D	▽					
5			◯	⇨	☐	D	▽					
6			◯	⇨	☐	D	▽					
7			◯	⇨	☐	D	▽					
8			◯	⇨	☐	D	▽					
9			◯	⇨	☐	D	▽					
10			◯	⇨	☐	D	▽					
11			◯	⇨	☐	D	▽					
12			◯	⇨	☐	D	▽					
13			◯	⇨	☐	D	▽					
14			◯	⇨	☐	D	▽					
15			◯	⇨	☐	D	▽					
16			◯	⇨	☐	D	▽					
17			◯	⇨	☐	D	▽					

Figure 4-17 Process Chart—Back of Blank Form

STEP	DETAILS OF ($\begin{smallmatrix}PRESENT\\PROPOSED\end{smallmatrix}$) METHOD	METHOD	OPERATION	TRANSPORT	INSPECTION	DELAY	STORAGE	DISTANCE IN FEET	QUANTITY	TIME (.0000)	TIME/HRS/UNIT	COST PER UNIT	TIME/COST CALCULATIONS
18			◯	⇨	☐	D	▽						
19			◯	⇨	☐	D	▽						
20			◯	⇨	☐	D	▽						
21			◯	⇨	☐	D	▽						
22			◯	⇨	☐	D	▽						
23			◯	⇨	☐	D	▽						
24			◯	⇨	☐	D	▽						
25			◯	⇨	☐	D	▽						
26			◯	⇨	☐	D	▽						
27			◯	⇨	☐	D	▽						
28			◯	⇨	☐	D	▽						
29			◯	⇨	☐	D	▽						
30			◯	⇨	☐	D	▽						
31			◯	⇨	☐	D	▽						
32			◯	⇨	☐	D	▽						
33			◯	⇨	☐	D	▽						
34			◯	⇨	☐	D	▽						
35			◯	⇨	☐	D	▽						
36			◯	⇨	☐	D	▽						
37			◯	⇨	☐	D	▽						
38			◯	⇨	☐	D	▽						
39			◯	⇨	☐	D	▽						
40			◯	⇨	☐	D	▽						
41			◯	⇨	☐	D	▽						
42			◯	⇨	☐	D	▽						

CONCLUSION

Figure 4-18 From-To Chart—Blank Form

FROM—TO CHART

DATE_____   PLANT_____
BY _____   PROJECT_____

ITEM(S) CHARTED_____   BASIS OF VALUES_____

OPERATION FROM \ OPERATION TO	1	2	3	4	5	6	7	8	9	10	11	12	13	14	15	16	17	18	19	20	TOTALS
1																					
2																					
3																					
4																					
5																					
6																					
7																					
8																					
9																					
10																					
11																					
12																					
13																					
14																					
15																					
16																					
17																					
18																					
19																					
20																					
TOTALS																					

QUESTIONS

1. Define a flow line.

2. What does flow analysis try to do?

3. What are the two basic groups of flow analysis techniques?

4. What are the fabrications of individual parts flow analysis techniques?

5. Draw a string diagram, multi-column process chart, and a from-to chart for the following four parts.

Part #	Weight	Sequence
A	1	1 2 3 4 7
B	2	1 3 2 6 7
C	3	1 3 4 5 6 7
D	4	1 3 4 5 7

What is the efficiency of the from-to chart?

6. Draw a process chart for tool box body shown in Figure 3-1 using the flow diagram in Figure 4-13.
7. What are the three techniques of total plant flow?
8. Draw an operations chart for your project.
9. Flow process charts combine what two techniques?

Chapter 5

Activity Relationship Analysis

Manufacturing flow has been discussed in Chapter 4, but there are other departments besides production which require good flow. Material flows from receiving, to stores, to warehousing, to shipping. Information flows between offices and the rest of the plant and people move from place to place. Each department, office and service faculty must be placed properly in relationships to each other. The techniques in this chapter:

1. activity relationship diagram;
2. worksheet;
3. dimensionless block diagram; and
4. flow analysis

will help us place each department, office, and service area in the proper place. Our objective is to satisfy as many important relationships as possible in order to create the most efficient layout possible.

The auxiliary services, personnel services, and offices discussed in this chapter will be discussed in much greater detail in Chapters 7, 8 and 11. The four techniques studied in this chapter are sequential. The activity relationship diagram is redrawn into a worksheet, and the worksheet is used to draw the dimensionless block diagram. The flow analysis is then drawn on the dimensionless block diagram.

THE ACTIVITY RELATIONSHIP DIAGRAM

The *activity relationship diagram* shows the relationship of every department, office, or service area with every other department and area. Closeness codes are used to reflect the importance of each relationship as follows:

Code	Definition
"A"	Absolutely necessary that these two departments be next to each other
"E"	Especially important
"I"	Important
"O"	Ordinary Importance
"U"	Unimportant
"X"	Closeness Undesirable

"A" codes should be restricted to movement of massive amounts of material. Great people movement could also be classified as "A" codes, but be sparing in the use of this most important code, otherwise it will become less useful.

"X" codes are as important as "A" codes but for the opposite reason. For example, if the Paint Department is located next to the Welding Department an explosion is possible.

Here is a step-by-step procedure for developing an activity relationship diagram:

1. List all departments in a vertical column on the right side of the form. For an example, see Figure 5-1.

2. Starting with line one (fabrication), establish the relationship code for each following department. To establish these relationship codes requires an understanding of all departments, an understanding of management attitudes, and a determination to produce the most efficient layout possible.

3. Reason codes can be used like asterisks. For example, we don't want shipping and receiving close to each other. Why? A 1 could be placed below the "X" $\frac{X}{1}$ in the 5, 8 intersection below the activity relationship code. We would write a reason code key below the diagram like:

Reason Code	Reason
1	For better flow
2	All material moves between these two departments
3	People movement
etc.	etc.

Figure 5-1 Activity Relationship Diagram

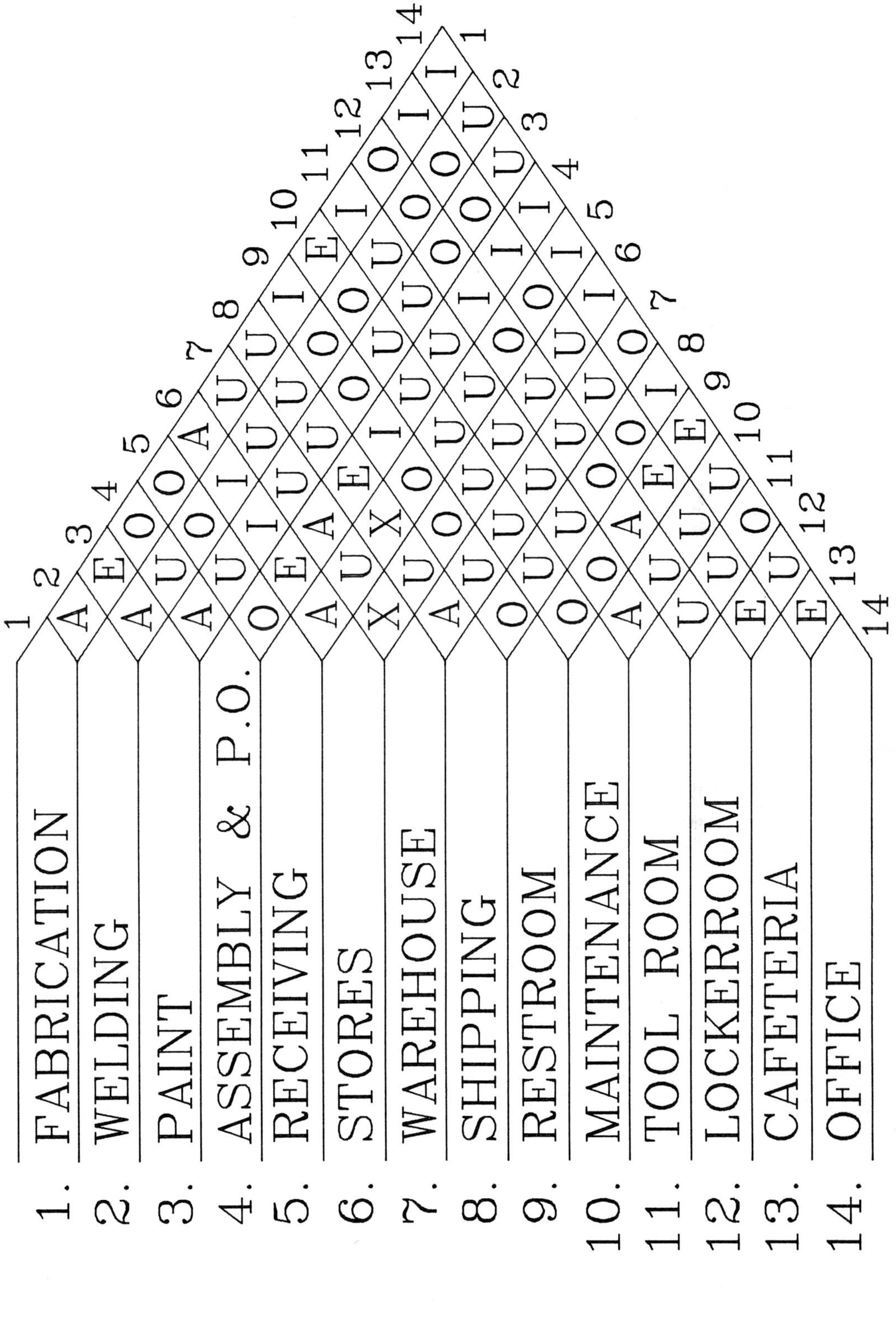

1. FABRICATION
2. WELDING
3. PAINT
4. ASSEMBLY & P.O.
5. RECEIVING
6. STORES
7. WAREHOUSE
8. SHIPPING
9. RESTROOM
10. MAINTENANCE
11. TOOL ROOM
12. LOCKERROOM
13. CAFETERIA
14. OFFICE

Next week someone may ask, "Why did you code this an 'A'?" Without reason codes you may not remember. Reason codes are not used all the time, but they often can be useful.

WORKSHEET

The *worksheet* is an interim step between the activity relationship diagram and the dimensionless block diagram. The worksheet will replace the activity relationship diagram. The worksheet interprets the activity relationship diagram and becomes the basic data for the dimensionless block diagram.

Here's a step-by-step procedure for the worksheet (see Figure 5-2):

1. List all the activities down the left side of a sheet of paper.
2. Make six columns to the right of the activity column and title these six columns, A, E, I, O, U, and X (relationship codes).
3. Taking one activity at a time, list the activity number(s) under the proper relationship code. Two points will assist the designer here:
   a) Be sure every activity number appears on each line (1–14 must be somewhere on each line).
   b) The activity codes for an activity listed below the number one activity on the activity relationship diagram are recorded above as well as below this activity: For example, line 2's (welding) relationship code with fabrication is an "A" and is located at the coordinate 1-2.

The activity relationship worksheet shows the same relationships as the activity relationship diagram.

Figure 5-2 Activity Relationship Worksheet

Activities	A	E	I	O	U	X
1. Fabrication	2, 6	3, 10	9, 11, 13, 14	4, 5, 12	7, 8	
2. Welding	1, 3		6	9, 10, 12, 13, 5	7, 8, 4, 11, 14	
3. Paint	2, 4	1	6	12, 13, 9	5, 7, 8, 10, 11, 14	
4. Assembly and P.O.	3, 7	6, 8	5, 9, 12, 13, 14	1, 5	2, 10, 11	
5. Receiving	6		14	4, 2, 1, 9, 12, 13	3, 7, 10, 11	8
6. Stores	5, 1	4	3, 2, 14	9	8, 10, 11, 12, 13	7
7. Warehouse	4, 8			14	5, 3, 2, 1, 9, 10, 11, 12, 13	6
8. Shipping	7	4	14	9, 12, 13	6, 3, 2, 1, 10, 11	5
9. Restrooms	12	13, 14	4, 1	8, 6, 5, 11, 3, 2, 10	7	
10. Maintenance	11	1		9, 2	8, 7, 6, 5, 4, 3, 12, 13, 14	
11. Tool Room	10		1	9, 14	8, 7, 6, 5, 4, 3, 2, 12, 13	
12. Locker Room	9	13	4	8, 5, 3, 2, 1	11, 10, 7, 6, 14	
13. Cafeteria		14, 12, 9	4, 1	8, 5, 3, 2	10, 8, 11, 7, 6	
14. Office		13, 9	8, 6, 5, 4, 1	11, 7	12, 10, 2, 3	

DIMENSIONLESS BLOCK DIAGRAM

The *dimensionless block diagram* is the first layout attempt. Even though this layout is dimensionless, it is the plan for the master layout and plot plan. Once size is determined, space will be allocated to each activity per the dimensionless block diagram's layout. If we obey the activity codes, a good layout will result. Sometimes it's harder laying out the dimensionless block diagram than when exact sizes are available because large departments tend to have more important relationships than small departments and can have many more departments (activities) close by them. Here's a step-by-step procedure for a dimensionless block diagram:

1. Tear up a sheet of paper into about 2″ × 2″ squares. (In our example, 14 squares are needed.)
2. Place an activity number in the center of each square.
3. Taking one square at a time, make a templet for that activity by placing the "A" relationship in the top left hand corner, the "E" relationship in the top right corner; the "I" relationship in the bottom left corner; the "O" relationships in the bottom right corner; "U" relationships are omitted, and the "X" relationships are placed in the center under the activity number. Each line on the worksheet will be one square.

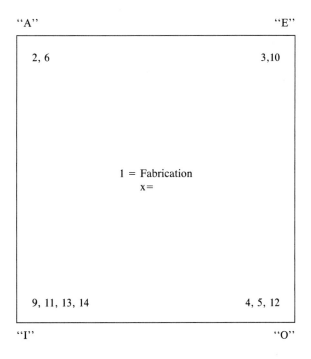

4. Now the 14 templets are ready to be placed in the best arrangement in order to satisfy as many closeness codes as possible. Start with the activity with the most important closeness codes. Place that templet in the middle of your desk. Look at the "A" codes, find those "A" coded activity templets, and place those services adjacent to the first templet with a full side against the first templet. In our example, pick up activity "I" (fabrication) and place it in the middle of your desk. Now pick up 2 and 6 (because they have "A" relationships with 1. Place templet 2 on any one of templet 1's four sides but have a full side contact. Place templet 6 also where it has a full side with one, but notice 6

Figure 5-3 Dimensionless Block Diagram

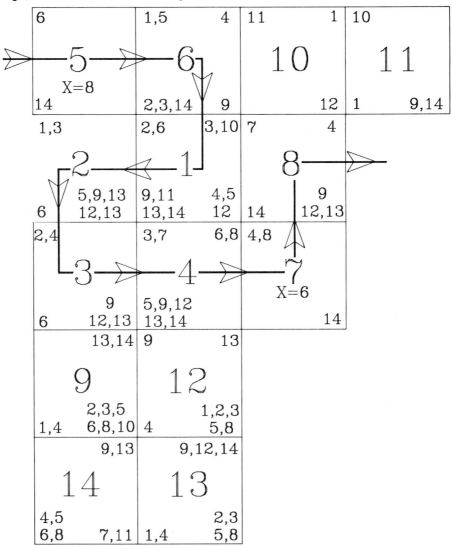

wants to be close (an "I" relationship with) to 2 also, so we allow a corner to touch. We now have three templets positioned. Continue to pick up additional "A" relationships and place where they satisfy the most activity relationships until all departments are accounted for. (See Figure 5-3.)

5. Grade your solution. All "A"'s must have a full side touching. All "E"'s must have at least a corner touching. No "X" relationship should be touching.

   Give two checkmarks for "A"'s not touching at all or for "X"'s touching with a full side. One checkmark for "A"'s with only a corner touching, with an "X" touching a corner or with an "E" not touching at least one corner.

How many checkmarks do you find? The fewer checkmarks the better. In our example:

$$6 - 4\sqrt{}$$
$$4 - 6\sqrt{}$$

FLOW ANALYSIS

Flow analysis is now performed on our dimensionless block diagram. Starting with receiving, show the movement of material to stores, to fabrication, to welding, to paint, to assembly and packout, to the warehouse, and to shipping. Flow analysis will insure important relationships will be maintained and that your layout makes good sense.

QUESTIONS

1. What are the six closeness codes and what do they stand for?
2. Where do these codes come from?
3. What is a reason code?
4. Develop a dimensionless block diagram for the activity relationship diagram in Figure 5-4.

Figure 5-4 Activity Relationship Diagram for Question # 4

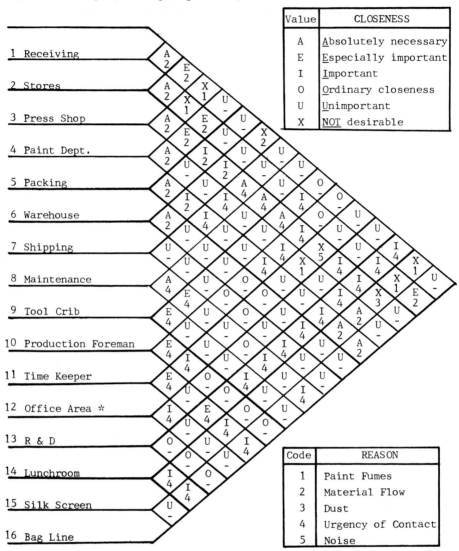

Value	CLOSENESS
A	Absolutely necessary
E	Especially important
I	Important
O	Ordinary closeness
U	Unimportant
X	NOT desirable

1 Receiving

2 Stores

3 Press Shop

4 Paint Dept.

5 Packing

6 Warehouse

7 Shipping

8 Maintenance

9 Tool Crib

10 Production Foreman

11 Time Keeper

12 Office Area *

13 R & D

14 Lunchroom

15 Silk Screen

16 Bag Line

Code	REASON
1	Paint Fumes
2	Material Flow
3	Dust
4	Urgency of Contact
5	Noise

* The Office Area has a seperate relationship chart.

CHAP. 5: ACTIVITY RELATIONSHIP ANALYSIS

Figure 5-5 Activity Relationship Diagram Blank Form

ENGINEER _____

DATE _____

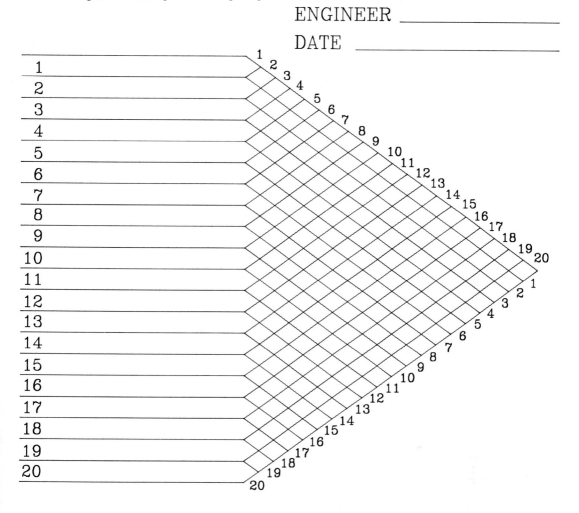

Figure 5-6 Worksheet For Activity Relationship Chart—Blank Form

Activity		*Degree of Closeness*					
		A	*E*	*I*	*O*	*U*	*X*
1.							
2.							
3.							
4.							
5.							
6.							
7.							
8.							
9.							
10.							
11.							
12.							
13.							
14.							
15.							
16.							
17.							
18.							
19.							
20.							

Figure 5-7 Dimensionless Block Diagram—Blank Form

A__ E__ X__ ___ I__ O__	A__ E__ X__ ___ I__ O__	A__ E__ X__ ___ I__ O__	A__ E__ X__ ___ I__ O__
A__ E__ X__ ___ I__ O__	A__ E__ X__ ___ I__ O__	A__ E__ X__ ___ I__ O__	A__ E__ X__ ___ I__ O__
A__ E__ X__ ___ I__ O__	A__ E__ X__ ___ I__ O__	A__ E__ X__ ___ I__ O__	A__ E__ X__ ___ I__ O__
A__ E__ X__ ___ I__ O__	A__ E__ X__ ___ I__ O__	A__ E__ X__ ___ I__ O__	A__ E__ X__ ___ I__ O__
A__ E__ X__ ___ I__ O__	A__ E__ X__ ___ I__ O__	A__ E__ X__ ___ I__ O__	A__ E__ X__ ___ I__ O__

_____ Chapter **6** _____

Workstation Design—Space Requirement

WORKSTATION DESIGN

The result of workstation design is a workstation layout and the workstation layout determines the space requirements. The manufacturing department's total space requirements are just a total of individual space requirement plus a contingency (a little extra) factor.

The workstation design is a drawing, normally a top view, of the workstation including the equipment, materials, and operator space. Designing workstations has been an activity performed by industrial engineers and technologists for nearly a century. During this period of time, the profession has developed a list of principles of *motion economy* that all new technologists should learn and apply. When the principles of motion economy are properly applied to the design of a workstation, the most efficient motion patterns will result.

"Where to start?" is the first question most often asked by new technologists faced with workstation design. The answer is very simple—anywhere! No matter where you start in workstation design, another idea will come along making that starting point obsolete. Where to start depends a great deal on what is to be accomplished at that workstation. The cheapest way to get into production is usually the best rule for the starting point. The cheapest way means just that—the simplest machines, equipment, and workstations. Any improvement on this cheapest method must be justified by savings. Therefore, the technologist is free to start anywhere, then improve on the first method.

The following information must be included in any workstation design:

1. Work table;
2. Incoming materials (materials packaging and quantity must be considered);
3. Outgoing material (finished product);
4. Operator space and access to equipment;
5. Location of waste and rejects;
6. Fixture and tools; and
7. Scale of drawing. (See Figure 6-1.)

A three-dimensional drawing would show an even greater amount of information. Any talented technician could attempt a three-dimensional design. Figure 6-2 is a photo of a well-planned workbench.

Figure 6-1 Workplace Layout—Old Method

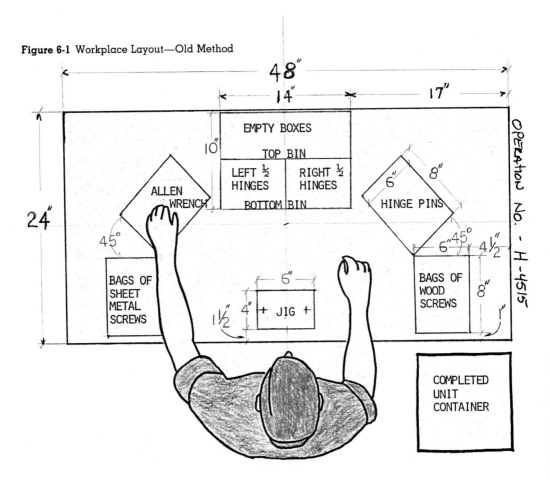

CHAP. 6: WORKSTATION DESIGN—SPACE REQUIREMENT

Figure 6-2 Workbench

Courtesy of American Seating Co.

The second example of workstation design will be of a machine operation (see Figure 6-3). The needs of this station design are the same as the previous station, but the equipment (machines, jigs, and fixtures) will be added.

Figures 6-4 through 6-9 are workstation designs for the equipment required in our tool box plant. Figure 6-10 is a workstation design for the tool box paint system.

Figure 6-3 Work Station Layout

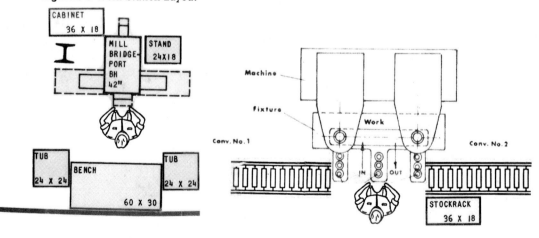

Figure 6-4 Strip Shear—Total Square Feet: 96

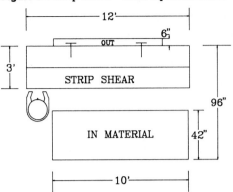

Figure 6-5 Punch Press—Total Square Feet: 88

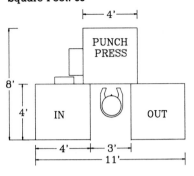

Figure 6-6 Chop Shear—Total Square Feet: 71.5

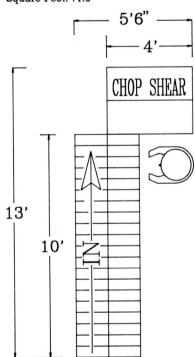

CHAP. 6: WORKSTATION DESIGN—SPACE REQUIREMENT

Figure 6-7 Press Brake—Total Square Feet: 88

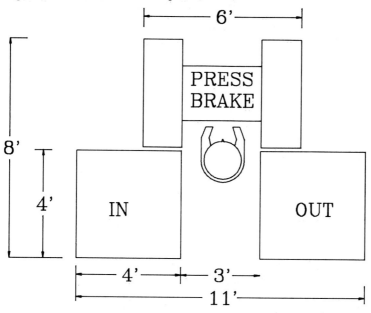

Figure 6-8 Roll Former—Total Square Feet: 108

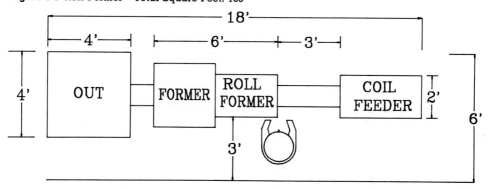

Figure 6-9 Fabrication Department Layout

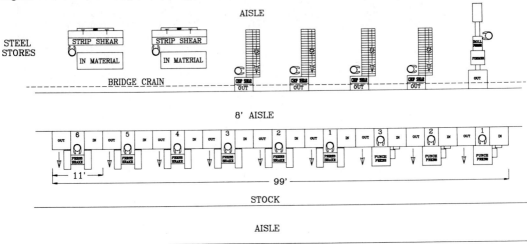

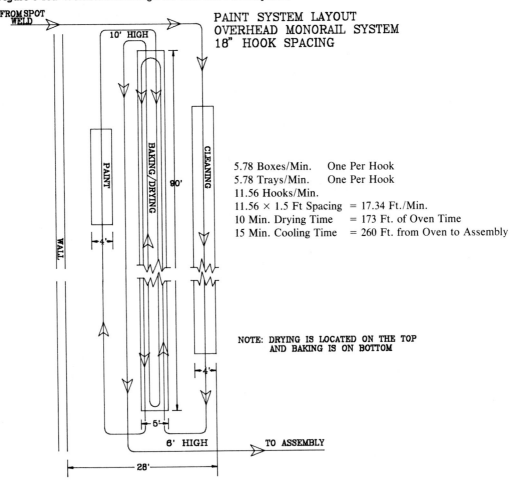

Figure 6-10α Workstation Design for Tool Box Paint System

PAINT SYSTEM LAYOUT
OVERHEAD MONORAIL SYSTEM
18" HOOK SPACING

5.78 Boxes/Min.   One Per Hook
5.78 Trays/Min.   One Per Hook
11.56 Hooks/Min.
11.56×1.5 Ft Spacing = 17.34 Ft./Min.
10 Min. Drying Time = 173 Ft. of Oven Time
15 Min. Cooling Time = 260 Ft. from Oven to Assembly

NOTE: DRYING IS LOCATED ON THE TOP
AND BAKING IS ON BOTTOM

Industrial engineers and technologists continually have been developing guidelines for more efficient and effective workstation design. Frank & Lillian Gilbreth originally collected these guidelines and titled them "The Principles of Motion Economy." Mr. Ralph Barnes has updated and published these principles since 1937.

Effectiveness is doing the right job. *Efficiency* is using the job right. Effectiveness is important to consider first because doing an unnecessary job is bad, but making a useless job efficient is the worst sin. Efficiency should be the goal of every industrial technology and especially of those involved with workstation design.

The principles of motion economy should be considered for every job. Sometimes, principles will be violated with good reasons. These violations and reasons should be written up for future use. You will have to defend yourself to every new technologist, so be prepared.

The principles are often used together in very creative ways. The only limit to improved workstation design is the technologists' creativity.

Principle 1. Hand Motions

The hands should operate as mirror images. They should start and stop motions at the same time, they should move in opposite directions, and they should both be working at all times.

If the hands are reaching for two parts at the same time, the bins should be placed equally back from the work area and the same distance from the center line of the workstation.

Reaching for only one part leaves the question of what the other hand is going to do. To keep both hands working at all times is a large challenge and can be most easily accomplished by doing two parts at a time (one with the left hand and one with the right). Holding parts in one hand while assembling other parts to it is a very poor use of the hands. It is said that the most expensive fixture in the world is the human hand.

Keep in mind that in workstation design, we don't consider people as being right-handed or left-handed unless hand tools are used. If hand tools are used, we consider everyone as being right-handed.

Principle 2. Basic Motion Types

Ballistic motions are fast motions created by putting one set of muscles in motion and not trying to stop those motions by using other muscles. Throwing a part in a tub and hitting a panic button on a machine are good examples. Ballistic motions should be encouraged.

Controlled or restricted motions are the opposite of ballistic motions and require more control especially at the end of the motion. Placing parts carefully on a belt is an example of a controlled motion. Safety and quality considerations are the best justification for controlled motions. But if ways of substituting ballistic motions for controlled motions can be found, cost reduction can result.

Continuous motions are curved motions and much more natural. When the body has to change direction, speed is reduced, and two separate motions result. If direction is changed less than 120°, two motions are required. Reaching into a box of parts laying flat on the table is an example requiring two motions: one motion to the lip of the box and another down into the box. If the box were placed on an angle, one motion could be used. This principle will be shown in greater detail in the gravity principle section of this chapter.

Principle 3. Location of Parts and Tools

Have a fixed place for all parts and tools and have everything as close to the point of use as possible. (See Figure 6-10b and 6-10c.) Having a fixed place for all parts and tools aids in habit-formation and speeds up the learning process. Have you ever needed a pair of scissors, and when you looked where they were supposed to be, they were gone. How efficient were you in the next few minutes? A tool makers tool box is layed out so that the tool maker knows where every tool is and can retrieve it without looking. That should be our goal in every workstation we design.

The need for having parts located as close to the point of use as possible is quite evident and, it should be of no surprise to anyone to learn that the farther you reach for something, the more costly that reach will be. Real creativity is required to minimize reaches. We can tier parts, instead of having one row of parts across the top of our workstation, or maybe three rows of parts one over the other would be better. We can hang tools from counterbalances over the workstation. Or we can use conveyors to move parts into and out of the workstation.

Principle 4. Release the Hands of as Much Work as Possible

As stated earlier, the hand is the most expensive fixture a designer could use. So we must provide other means of holding parts. Fixtures and jigs are designed to hold parts so that we can use both hands. Foot operated control devices can be designed to operate equipment to relieve the hands of work. Conveyors can move parts past operators for their work so they don't have to get or aside the base unit. Powered round tables are also used to move parts past an operator. (See Figure 6-10d.)

Fixtures can be electric, air, hydraulic, and manually operated. They can be clamped with little pressure or tons of pressure. They can have any shape which is

Figure 6-10b Location of Parts and Tools

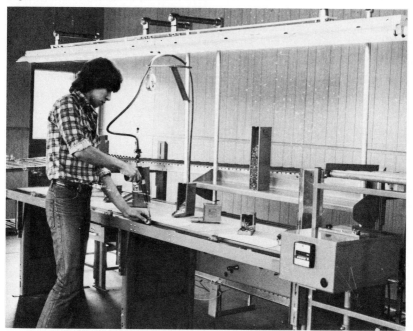

Courtesy of Alden Systems Co., Inc.

Figure 6-10c Location of Parts and Tools

Courtesy of Alden Systems Co., Inc.

CHAP. 6: WORKSTATION DESIGN—SPACE REQUIREMENT

Figure 6-10d Release the Hands of as Much Work as Possible

Courtesy of Alden Systems Co., Inc.

Figure 6-10e Operator Considerations

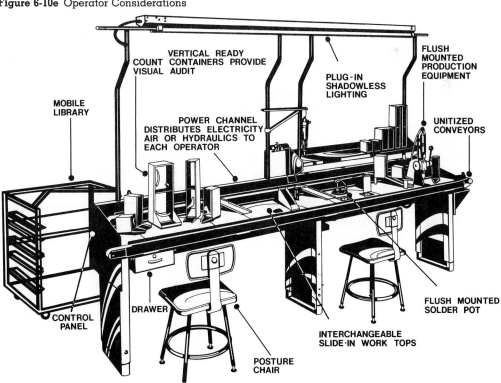

VERTICAL READY
COUNT CONTAINERS PROVIDE
VISUAL AUDIT

PLUG-IN
SHADOWLESS
LIGHTING

FLUSH
MOUNTED
PRODUCTION
EQUIPMENT

MOBILE
LIBRARY

POWER CHANNEL
DISTRIBUTES ELECTRICITY
AIR OR HYDRAULICS TO
EACH OPERATOR

UNITIZED
CONVEYORS

FLUSH MOUNTED
SOLDER POT

CONTROL
PANEL

DRAWER

INTERCHANGEABLE
SLIDE-IN WORK TOPS

POSTURE
CHAIR

Courtesy of Alden Systems Co., Inc.

dictated by the part. A hex nut can be placed in a hex-shaped hole that has no clamping need, but will be held firm because of the part and fixture shape. Fixture design is easy and only your knowledge of the part and needed processes are required to design fixtures. Many tooling vendors would love to supply you with fixture building materials.

Principle 5. Use Gravity

Gravity is free power. Use it! Gravity can move parts closer to the operator. By putting an incline in the bottom of parts hoppers, parts are moved closer to the front of the hopper. Production management loves for us to spare every expense, and the use of gravity can do that.

For example, consider a box which is 24″ × 12″ × 6″ laying flat on the table. The average part in that box (the only part the designer is interested in) is 12″ back, 6″ over, and 3″ down the exact middle of the box. Now if we get a scrap 2″ × 4″ board out of the trash and place it under the rear end of the box and raise it up 4″ to 5″ inches, the parts will slide down to the front of the box as the parts are used. The operator's reach has been reduced from 12″ to 3″ from the front lip of the box—a significant cost reduction.

Large boxes of parts can be moved into and out of workstations using gravity rollers and skate wheels. Parts can be moved between workstations on gravity slides made of sheet metal, plastic, and even wood.

Gravity can also be used to remove finished parts from the workstation. Dropping parts into chutes that carry the parts back, down, and away from the workstation can save time and workstation space. Slide chutes can carry punch press parts away from the die without operator assistance by using jet blasts, mechanical wipers, or the next part pushing the finished part from the die.

Gravity use is everywhere. The workstation designer should try to incorporate as much gravity use in his or her design as possible. Designing the use of gravity into your workstation designs is fun. Opportunities are everywhere. Find them!

Principle 6. Operator Considerations

Efficient operators must be allowed to work at the right height, given comfortable chairs, with enough light and adequate space to perform their tasks.

The *correct work height* is elbow height. With the forearm held parallel to the ground and the upper arm straight down, measure the elbow height to the floor. This is the work height. A job should be designed for sitting or standing, but the elbow height must be the same. This requires the designer to calculate working height while standing, then provide a chair that will accommodate that height while sitting.

The chair will have to be *adjustable*. Since work height is dependent on the individual, chairs and tables will have to be adjustable for efficient operation. The chair must also be comfortable. This usually means that it supports the back. Also, a foot ring helps comfort. Comfortable chairs and the option of sitting or standing give the operator a chance to recuperate while working. The end result is more output and less fatigue.

Adequate lighting may not be available in the normal lighting of a manufacturing department, so additional lighting would be added—much like a desk lamp. Where to place this lighting is the problem. The best place is over the work and slightly over the back, but not casting a shadow. Much lighting is placed in front of the work, but this causes glare from the reflection. Auxiliary lights could be placed to the left or right of the work as well.

Operator space should be 3′ × 3′ unless the workstation is wider, but 3′ × 3′ is normal. Three feet off the aisle is adequate for safety, and 3′ from side to side allows parts to be placed comfortably next to the operator. If two people are working back to back, then 5′ between stations is recommended. If machines need maintenance and clean-up, a 2′ access should be allowed around the machine. Movable equipment can be placed in this access area if needed for efficient operation.

SPACE DETERMINATION

The *space determination procedure* for most production department starts with workstation design. From each workstation layout, we measure the length and the width to determine the square footage of each station. The following data resulted from the workstation layouts in Figures 6-4 to 10, 3-7 and 3-8.

	Length	×	Width	=	Sq. Ft.	×	# Stations	Total Square Ft.
Strip Shear	12	×	8.5		102		2	204
Chop Shear	15	×	5		75		4	300
Punch Press	11	×	8		88		3	264
Press Brake	11	×	8		88		6	528
Roll Former	17	×	6		102		1	102
Paint System	100	×	28		2,800		1	2,800
Spot Welding	34	×	28		952		1	952 (Fig. 3-7)
Assembly	38	×	16		608		1	608 (Fig. 3-8)
	Total Square Feet							5,758
	Times 150% =							8,637

Multiplying the total square feet by 150 percent allows extra space for the aisle, work in progress, and a small amount of miscellaneous extra room. It does not include restrooms, lunch rooms, first aid, tool rooms, maintenance, offices, stores, warehouse, shipping, or receiving. These area requirements will be discussed in Chapters 7 and 8. The extra 50 percent space added to the equipment space requirement will be mostly used for aisles. Aisles can be very space consuming, for example, lets lay out a 100′ × 100′ plant as follows:

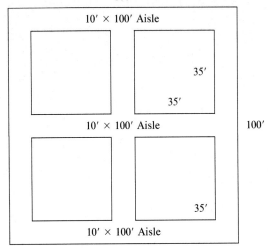

100′

10′ × 100′ Aisle

35′

35′

10′ × 100′ Aisle

100′

35′

10′ × 100′ Aisle

A 10′ aisle around the outside of the production area will eliminate clutter next to the walls. But that leaves us with an 80′ × 80′ area with no aisles, so lets put in 10′ cross aisles. How much room did we use?

(3) 100′ long, 10′ wide aisles	=	3,000 ft.^{2}
(3) 70′ long, 10′ wide aisles	=	2,100 ft.^{2}
Total Aisle Square Feet		5,100 ft.
Total Square Feet (100′ × 100′)		10,000 ft.^{2}

$$\frac{5,100 \text{ ft.}^2}{10,000 \text{ ft.}^2} = 51\% \text{ Aisles}$$

Our 50 percent extra space would not be half enough for this kind of aisle layout. We would have to add lots more space for a plant with 50 percent of its space taken up by aisles. A better aisle plan might be similar to the one shown at the top of page 97.

Two 100′ aisles, 8′ wide equals 1,600 ft.^{2}

$$\frac{1,600 \text{ ft.}^2}{10,000 \text{ ft.}^2} = 16\% \text{ Aisles}$$

This may be too light, but notice the improvement from 51 percent down to 16 percent and we have better access to areas (35′ vs. 28′ wide areas).

Small space consuming items such as an air compressor or drinking fountain may be included in this 50 percent extra area, but large area requirements must be designed and planned. The next chapter addresses those other areas that require space designs.

Figure 6-11 is a blank form for your use. Figure 6-12 illustrates commercially available workstations.

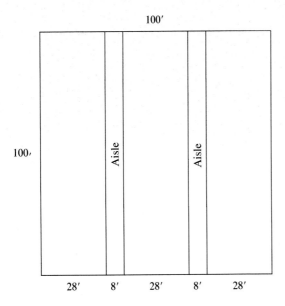

100'

100'

28' 8' 28' 8' 28'

Aisle   Aisle

Figure 6-11 Machinery and Equipment Layout Data Sheet—Blank Form

FRED MEYERS & ASSOCIATES	MACHINERY & EQUIPMENT LAYOUT DATA SHEET

DESCRIPTION OF MACHINERY & EQUIPMENT: DATE:

DESCRIPTION OF OPERATION:

DESIGNED BY: COMPANY NAME: LOCATION:

PHOTO   LAYOUT DRAWING

SCALE =

Reference Notations/Changes

MACHINE SPECIFICATIONS

Figure 6-12 Workstation

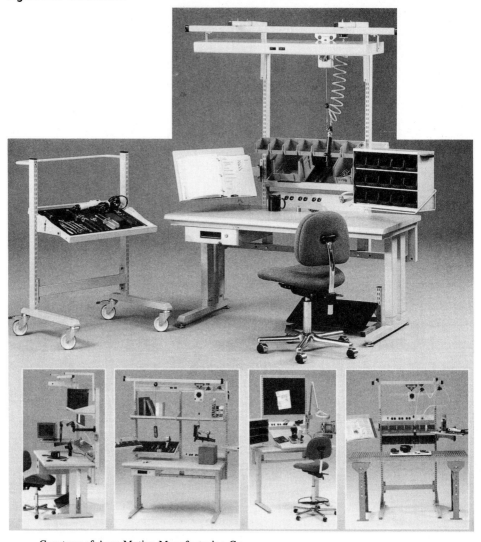

Courtesy of Aero-Motive Manufacturing Co.

Ergonomics is the study of workplace design and the integration of the worker with his environment. Bringing the worker and the task to be performed together in the most efficient way possible is our goal. Ergonomics is an important subject in industry today. Figure 6-13 is a write-up courtesy of Aero-Motive Manufacturing Company which describes the importance of Ergonomics.

CHAP. 6: WORKSTATION DESIGN—SPACE REQUIREMENT

Figure 6-13 Ergomation

ERGOMATION is the successful integration of the worker and the process environment . . . bringing together human and mechanical elements in the most efficient way possible to increase productivity and protect your investment in both worker and equipment.

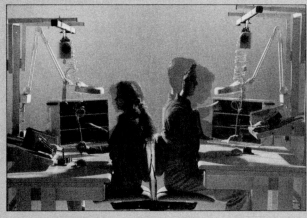

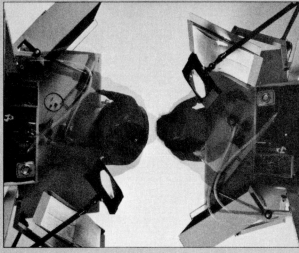

ERGONOMICS has become a critical issue in today's workplace. Research indicates that thousands of work hours are lost each year as a result of RMI (Repetitive Motion Injury) and related musculoskeletal conditions attributed to repetitive assembly tasks. This factor, combined with escalating workmans' compensation costs, illustrates the need for work systems that minimize the threat of RMI as well as optimize productivity. The Aero-Motive Workstation System is the answer.

FLEXIBILITY is important to everyone involved in the work process. Ergomation Products are designed to allow for operator differences in height, weight, reach and strength. They provide workers with infinite adjustment options for their work surface, chairs, component bins, tools, power, lighting and any other work-related accessories. The upright system allows for optimum placement of all components—via articulating arms and a shelving system designed to serve the first and second horizontal and vertical reach zones. This creates a work envelope that reduces unnecessary movement and places the worker in the best possible position for the tasks at hand.

MODULARITY is an essential advantage of the system. All components—from uprights, to work surface to attachments—feature universal connecting hardware that permits a limitless variety of configurations. Carts can be equipped with shelves, containers and articulating arms to facilitate kitting and in-line transport. Power supplies provide for air, vacuum, nitrogen, electricity or communication lines, and can be rearranged easily.

Look inside, and see the advantages of Ergomation at work for yourself.

Courtesy of Aero-Motive Manufacturing Co.

QUESTIONS

1. Where do you start on workstation design? Why?
2. What is the starting point of workstation design? Why?
3. What must be included in a workstation design?
4. What are the principles of motion economy?
5. What is effectiveness?
6. What is efficiency?
7. What is the extra 50 percent space added to the workstation space requirement?
8. Design workstations for your project and develop the fabrication area requirement in square feet.

Chapter 7.

Auxiliary Services—Space Requirement

Manufacturing departments need support services, and these services need space. The purpose of this chapter is to identify these services, define the purposes of these services, determine the facilities requirements, and determine the space requirements. There are many service functions to consider in a manufacturing plant, but the biggest area consumers are:

1. receiving and shipping;
2. storage;
3. warehousing;
4. maintenance and tool room; and
5. utilities, heating, and air conditioning.

RECEIVING AND SHIPPING

Receiving and shipping are two separate departments, but have very similar people, equipment, and space requirements. Receiving and shipping could be placed next to each other or across the plant from each other. The placement of the receiving and shipping department has a big affect on the flow of material in the plant. The receiving department is the start of the material flow, while the shipping department is the end of the material flow.

The Advantages and Disadvantages of Centralized Receiving and Shipping

A centralized receiving and shipping point would have the following advantages:

1. common equipment;
2. common personnel;
3. improved space utilization; and
4. reduced facility costs (less outside construction costs).

Loading and unloading trucks are very similar functions, so the facilities are similar. Dock doors, dock plates, fork trucks, and aisles are needed for both receiving and shipping. In some plants, it could be the same dock. Personnel requirements are also similar. Responsible people who know the value of proper counts, proper identification, and control of the company's most valuable assets are receiving and shipping clerks.

The disadvantages of centralized shipping and receiving are space congestion and material flow. Space congestion can cause injury, product damage, and lost materials. It would be a costly mistake to ship out some of newly received parts. Material flow is more efficient if the material could flow straight through the plant: receiving on one side of the plant and shipping on the other side.

Receiving in more than one place is also a possibility. Steel plate could come into the plant via its own area, finished parts could enter the plant close to assembly, while all other raw material come in a third receiving area. The most cost efficient method is the correct choice.

Choosing to place shipping and receiving close together or across the plant from each other is a difficult decision based on balancing the advantages and the disadvantages. The result will be an activity code of ''A'' or ''X.'' You and your management will have to choose and that choice will dictate the flow of material through your plant.

The Trucking Industry's Affect on Shipping and Receiving

The trucking industry can affect receiving and shipping departments. The trucking industry is organized nationally to deliver raw materials and parts to industry in the morning, and pick up shipment in the afternoon. This is known as *less than truck load quantities* (LTL). Full truck loads are handled differently, but if you look at the sources of our raw materials, it could come from hundreds of sources. No one would expect a truck to show up at our dock with one box of parts, and a full truck load could be years worth of inventory, so we use common carriers. A truck arrives in our town with many orders for many plants. That truck and many more are unloaded at a local trucking company's warehouse. The materials are sorted by company to be delivered the next morning. Overnight, the local trucks are loaded for delivery. Several plants could be loaded on the same truck with the first stop loaded last and last stop loaded first. The truck stops by our receiving department and drops off our many raw materials and parts orders for the day. In the afternoon, the same truck could return and pick up our shipments. One truck could pick up 50,000 pounds of shipment, take it to their break bulk station and separate into destinations for interstate trucks to pick up on their way through our town.

The Functions of a Receiving Department

The functions of a receiving department are to:

1. receive trailers;
2. unload;
3. record receipt;
4. open, separate, inspect and count;
5. over shortage and damage reports;
6. receiving reports; and
7. send to stores or production.

Receiving trailers. Trailers are backed up to the receiving dock doors, the tires are chalked, the trailer doors are opened, a dock board or dock plate is positioned between the trailer and the floor of the plant, and the driver gives the receiving clerk a manifest which tells the receiving clerk what to unload.

Unloading. The material is removed from the trailer and placed on the dock in the holding area. The receiving clerk signs the trucker's manifest (acknowledging the receipt of so many containers) and the trucker leaves. No count of material or quality check need take place before the driver leaves, but visible carton damage should be noted on the driver's paperwork.

Record receipt. When material is unloaded, it is checked in on a log. This log is often called a *Bates log* after the name of a sequencing number stamp called a *Bates stamp*. The Bates stamp has the ability to stamp the same number 3 times before advancing to the next number. This number is stamped on the Bates log, the parking slip, and the receiving report. The Bates log is simply the sequential record of the truck's receipt. Starting with the Julian calendar date (a three-digit number indicating the day of the year) the next three digits are the order that trucks came in today. For example, July 3rd is the 185th day of the year and this is our 21st truck arriving today, so:

Bates Number	Trucking Company	# Containers
185021	Arkansas Best Freight	15
185022	Allied	4

Open, separate, inspect, and count. During the first hours of the day, there may not have been time to open a single container to officially check in the merchandise, but before the day is complete, everything received today must be opened, separated, inspected, and counted. Opening each container to check what's inside is a must. The first check is to make sure that everything in the container is the same part number. If not the same item, we must separate and categorize each part number so that it can be stored separately. After separation, a quality check must be made to see if this is what we ordered. Strength of material as well as visual inspection may be needed. In this case, the Quality Control Department may have a large facility requirement for the receiving area. The quantity must also be checked. If the vendor (supplier) said they shipped 10,000 and we didn't count, we could pay for parts never received.

Over Shortage and Damage reports (OS&D). If the count is either over or under, an OS&D report is prepared and sent to purchasing for resolution. Damage suffered in shipment and quality problems are also reported on this form. Each problem becomes a project for the purchasing department who has to work it out with the supplier, but the eyes and ears of our company is with our receiving department. It is said that receiving is the key to the company's bank, because sloppy receiving can give away thousands of dollars.

Receiving reports. The receiving report is the notice to the rest of our company that a product has been received. Our supplier receives our purchase order for some of their products. They in turn create a shipper, fill the order, and attach a copy of their shipping order to a box of our order. This is called *packing list*. At almost the same time that the product is shipped to us, an invoice (bill) is sent in the mail. When we receive the shipment, we find the container with the packing list in it. Some companies use the customer's packing list for a receiving report, but it's better to have our own uniform report for checking things in, and we also have a record of the receipt. After we've checked quality and quantity, we issue the receiving report to accounting. The accounting department (accounts payable) collects copies of the purchase order, receiving report and invoice. Only after all three documents are received is the bill paid. You see we will pay for what receiving said we received. Errors can be very costly at receiving. The receiving report will carry the following information:

1. P.O. number (purchase order number)
2. Vendor's name and address
3. Date
4. Part number(s)
5. Part name(s)
6. Quantity
7. Bates log number
8. Packing list number

Send to stores or production. Once all receiving functions are complete, the product is set in an area between receiving and stores awaiting disposition to production stores or to production operations depending upon urgency. This is the holding area awaiting fork truck drivers to move the material off the receiving dock.

Facilities Required for Receiving Departments

Dock doors, dock plates, aisles, outside parking lots, maneuvering space, roadways, and offices are a few examples of facilities needed in receiving departments. The number and size of these facilities depend upon our product, its size, and quantity produced.

Dock doors. The number of dock doors needed is dependent upon the arrival rate (trucks per hour) at peak time, and the service rate (unloading time). For example, if 12 trucks arrive during a peak hour, and it takes 15 minutes to unload an average truck, three dock doors would be needed. Fifteen minutes per truck would allow us to unload four trucks per hour per door, so three doors would be needed. This is the simple arithmetic way of figuring.

Dock plates, dock levelers, and dock boards. These are all tools used to bridge the gap between the floors of buildings and the floors of trailers so that material can be moved on and off the trailer easily. There is a big difference in the cost of these facilities. They will be further discussed in Chapter 10 on material handling.

Aisles. Aisles leading from the trailers into the plant must be sized for the material handling equipment, the material being moved, and the frequency of trips. Generally, aisles into trailers are 8' wide because that is the width of a trailer, but sometimes a trailer is unloaded from the side or with overhead bridge cranes. Plan for such differences.

Outside areas. The area around the outside of the loading dock should be well-planned (see Figure 7-1). Space considerations should take into account that:

1. Trailer parking alone can take up 65' out from the plant wall.
2. Maneuvering space is the space between the road and parking area and is usually about 45'.
3. Roadways are 11' one way or 22' for two-way traffic.

Offices. Offices on the receiving dock are normally very small. Space for a desk, files for purchase orders, Bates logs, receiving reports, and over shortage and damage reports are all that is needed. Depending upon the number of people assigned to the receiving area, 100 square feet per clerk is needed.

Figure 7-1 Receiving Area

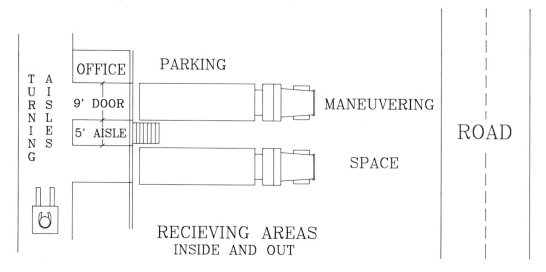

RECIEVING AREAS
INSIDE AND OUT

Space Requirements for the Receiving Department

The first method of determining receiving dock space calls for visualization of the receiving job based on the number of finished products produced per day and the weight of those units. For example, if we are making 2,000 tool boxes per day and those tool boxes weigh five pounds each, 10,000 pounds of steel will be required every day. So, on the average, 10,000 pounds will be received and shipped every day. Some days it will be 5,000 pounds, other days 15,000 pounds, but on the average 10,000 pounds per day. Our receiving dock will be sized to receive 10,000 pounds. What does 10,000 pounds of steel look like? Consider that 40,000 pounds is a truck load. We need only one-fourth of a truck load space. A semi-trailer is 8′ × 40′ long and steel would be only stacked a few feet high, so 10,000 pounds would be one-fourth of 8′ × 40′ or 80 square feet. Multiply this by two to allow for aisles, office, etc. and our dock is 160 square feet, about 12′ × 13′, a very small area which could have only one door. The outside area for parking is extra. Figures 7-2 and 7-3 are examples of receiving departments space requirements.

The second method of receiving department space determination is the *facility approach*. We'll need the following data:

1. dock doors;
2. aisles;
3. unloading hold area;
4. working area to open, separate, count and check quality;
5. office area; and
6. holding area for stores.

Figure 7-2 Receiving Dock for Steel

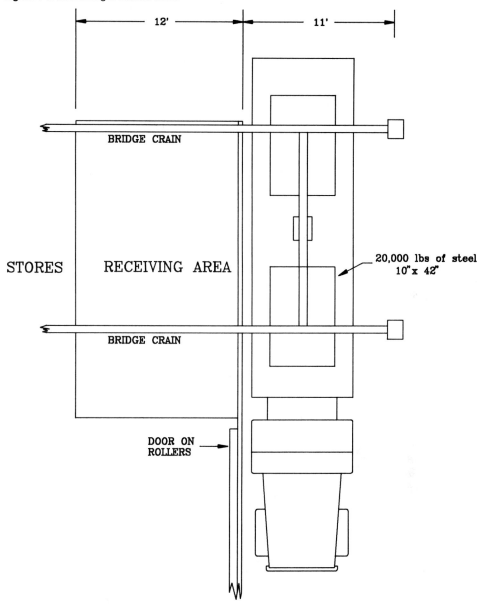

Figure 7-3 The Exterior of the Receiving Dock

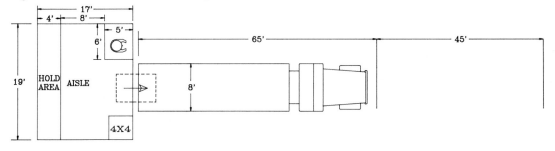

The holding area would still be 10,000 pounds worth and could be slightly larger for a work area that would move with progress through the stack of holding area. The office area is 100' per person (in this case no more than one person).

The Functions of a Shipping Department

The functions of the shipping department include:

1. Packaging finished goods for shipping.
2. Addressing cartons or containers.
3. Weighing each container.
4. Collecting orders for shipping (stage).
5. Spotting trailers.
6. Loading trailers.
7. Creating bills of lading.

Packaging finished goods for shipping. This process varies with the product and the kind of company. One company may have thousands of products with one customer ordering a few hundred items. These items are pulled together and packaged. The package may be a box or a pallet or even a cargo container. Let's consider a hand tool company. They would pack their orders in heavy duty cardboard boxes. Packaging must include careful placement of individual items so they are not damaged in shipment. This may require wrapping, stuffing, nesting, and even specially designed shock absorption material. The weight of the container must be compatible with our customers' ability to unload the shipment. Packaging workstation design must also consider the principles of motion economy. Proper work height, good lighting, all tools and materials located conveniently are only a few of the principles of motion economy that must be considered. See Figure 7-4 for a typical packaging station.

Addressing cartons or containers. This is required if the order goes by common carrier (LTL). Just like a letter, the order (boxes) go into a system with many other orders. Each box must be addressed. When many boxes are going to the same customer, a stencil may be produced to mass produce the address. Some systems have computer-generated shipping labels and others use a copy of the shipper as an address label. The important fact is that every container must be addressed. Efficiency (or cost reduction) will determine which addressing system to use.

Weighing each container. This process is required for several reasons. First, the trucking company will charge us by the pound, so we need to know the weight to determine trucking costs. Second, a quality control technique is used to compare the weight of each order to the individual weight of each part shipped. If the container doesn't weigh enough, something was left out. If the container weighs

Figure 7-4 Packaging Workstation

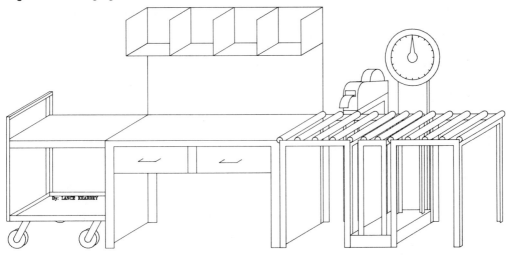

too much, something extra was placed in the box. When our customers receive the shipment and claims a shortage, we can check the weight to verify the shortage. If the weight checks out, we ask the customer what they got instead because the weight was correct. Third, trucks can haul only specific maximum weights. We must insure we don't overload them. Lastly, we can use the weight as an output figure for productivity calculations. In a warehouse, the pounds shipped can be divided by the hours worked to create a performance indicator of pounds shipped per man hour. We should be getting better every year and pounds shipped per man hour is a good indication of performance.

Weight scales can be built into the conveyor line as shown in Figure 7-4, or the scale could be built into the floor so fork trucks can weigh whole pallets. (See Figures 7-5 and 7-6.)

Collecting orders for shipping. This is often called *staging orders*. Our company may use four trucking companies to move all our freight: one for all freight going north, another trucking company may take all our freight going west, a third company goes south, and the final company going east. All day long as we fill and pack orders, we place the finished packaging in the proper staging area for that truck line.

Spot trailers. The trucking company sends in a trailer in the afternoon to pick up our freight. This is called spotting a trailer. Some big shippers may talk the trucking company into leaving a trailer at our plant all day. Then we can stage the shipments on the trailer and save plant space.

Loading trailers. Loading the trailer can be done very quickly if pallets are used. Most trailers will hold a maximum of 18 pallets.

Figure 7-5 Weight Scale—Drive On

Capacities to 6,500 lbs/3,000 kg

Semi-frame

Full Frame

Courtesy of Toledo Scales.

Bills of lading. As the trailer is loaded, the bill of lading is created. The bill of lading lists every order and the weight of that product. The bill of lading is the truck driver's authorization to remove the product from the plant, and will eventually come back to us as a bill for the trucking service.

Courtesy of Hytrol Conveyor Co.

Space Requirements for Shipping Departments

Space for shipping must include areas for packaging, staging, aisles, trailer parking, roadways, and offices. Sometimes, lounges for truckers and restrooms are included. As in the receiving department, the overall weight of our shipment will help us visualize the size of our daily shipments. Two thousand tool boxes per day times five pounds per tool box equals 10,000 pounds per day. But there is a lot of air in a tool box, so how many cubic feet does 2,000 tool boxes take up?

$$\frac{8'' \times 8'' \times 18''}{1{,}728 \text{ cubic in./ft.}} = .66 \text{ cubic feet} \times 2{,}000 = 1{,}333 \text{ cubic feet/day}$$

A trailer is 8′ wide × 40′ long × 7′ high or 2,240 cubic feet.

$$\frac{1{,}333 \text{ cubic feet required}}{2{,}240 \text{ cubic feet/trailer}} = .6 \text{ trailer/day}$$

One dock door is required. Space to store (stage) a days supply of shipments (1,333 cubic feet) is required. A space of 8′ × 40′ × 60 percent = 192 square feet for staging. Multiplying this times 200 percent will put in the extra space needed for aisles and offices, but not packaging. Packaging is based on workstation layout (like production) but our tool box example doesn't need much packaging—just

Figure 7-7 Square Footage of Shipping Department

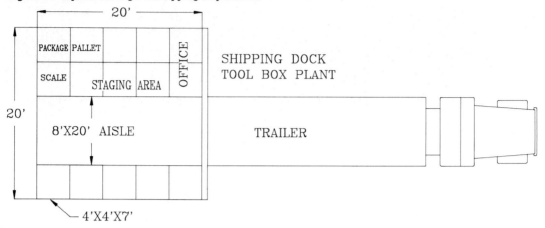

addressing and weighing. The tool box plant's shipping department will be about 400 square feet inside the plant plus parking for one trailer. See example in Figure 7-7 for the plant's shipping department.

STORAGE

Stores is a term used to denote an area set aside to hold raw materials, parts, and supplies. There are many different types of stores:

- Raw material stores;
- Finished parts stores;
- Office supplies stores;
- Maintenance supplies stores; and
- Janitorial supply stores.

Each of these stores requires space and must be considered when calculating total space requirements, but raw material stores and finished parts stores are the biggest users of space. Our primary interest will be raw material stores, but the same procedure can be used in calculating space for other stores.

The space requirements for stores is dependent upon the stated inventory policy of the company. The policy could be as straight forward as, "Provide space to store a one-month supply of raw material," or a little more creative policy might be, "Provide an area to store a one-weeks supply of "A" items, 2 weeks of "B" items, and a one-month supply of "C" items." "A" items are those parts that account for 80 percent of the inventory value. Usually 20 percent of the part numbers make up 80 percent of the dollar value. In an automobile assembly plant, the engine/transmission is the most expensive part of the automobile and can cost

$2,000 of the $12,000 total cost (or 17 percent). There may be over 2,000 parts to any car.

Inventory Classification	Percentage of Parts	Percentage of $	Inventory Policy
A	20	80	One-week Supply
B	20	15	Two-weeks Supply
C	60	5	One-month Supply

In our tool box example, if we make 2,000 tool boxes per day at a material cost of $5 each and inventory a 20-day supply, we would have $200,000 in inventory. A carrying cost of 25 percent per year is normal, so the cost of carrying a one-month supply of inventory is $50,000 per year. If we redesigned our system and instituted an ABC inventory system we would reduce costs to 25 percent carrying cost × $63,000 = $15,750/year carrying cost.

A	80% for 1 week	$40,000*
B	15% for 2 weeks	15,000
C	5% for 1 month	8,000
Total Inventory Value		$63,000

* 80% of $5.00/unit × 2,000 boxes
× 5 days = $40,000

We have saved $34,250 in inventory carrying cost. The less inventory you carry, the lower the cost if you don't run out of material. Large inventory allows production management to be very comfortable. They don't need to worry about running out of material as often, but at what cost? Carrying cost measures the cost of carrying inventory. The 25 percent includes:

1. Interest for borrowing the money to buy the inventory of raw material (say about 12 percent).
2. The space for storing, heating, cooling, lighting the material (let's say about 8 percent).
3. Taxes, insurance, damage, obsolescence, etc. (about 5 percent).

These costs are real costs that add no value to our product.

The cost of running out of a single item of inventory used on our production line could stop the whole plant, so some inventory is needed. How much inventory is a management decision? Looking at the "A" item again, 20 percent of the part numbers account for 80 percent of the inventory cost. The philosophy is that the less you have of this most expensive class, the better. But you will need to reorder it four times as often as a "C" item. This also means four shipments, four receivings, four orders, and so on, so ordering cost will increase, but only on these most important 20 percent of the part numbers,

Just in Time Inventories

Just in Time (JIT) is the inventory policy that has been made famous in Japan. Primary manufacturers depend on their suppliers to deliver parts as often as every four hours, thereby eliminating the need for raw material inventory storage area. JIT depends upon unfailing vendor performance. Vendors at far distances from our plant would have to warehouse their product in our area. This is a very special inventory subject that takes total corporate commitment and very special relationships with vendors. JIT will affect the plant layout in many ways:

1. Adjust or eliminate receiving, receiving reports, and so on.
2. Eliminate incoming quality control checks.
3. Eliminate or greatly reduce stores area requirements.

In this text, we will not consider JIT because designing a layout for a non-JIT system is more difficult and, unfortunately, more common.

The goals of any stores department should be to:

1. Maximize the use of the cubic space.
2. Provide immediate access to everything.
3. Provide for the safekeeping of the inventory including:
   a. damage; and
   b. count control.

Maximizing the Use of the Cubic Space

Maximizing the use of the cubic space requires the use of racks, shelves, mezzanines, and minimizing aisle space and empty space. This brings us to the number one design criteria for a store room.

Leave room to store only half the required inventory.

To explain this design criteria, an inventory graph is needed (see Figure 7-8). Inventory graph terms include:

1. *Units on hand* is the y axis (vertical axis) and it measures how many units of this part number remain in the inventory.
2. *Days* is the x axis (horizontal axis) and measures the day of the year this day represents. In the life of a product, this axis could be very long, but a years worth of data would be very useful.

Figure 7-8 Inventory Curve—One Curve Per Part Number

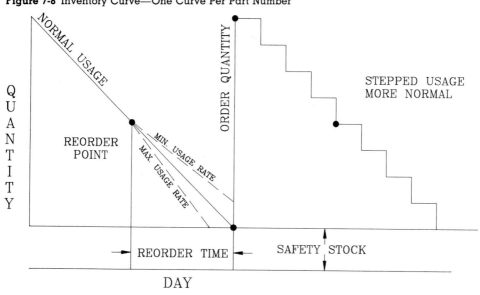

3. *Order quantity* is how many units we order at a time. If we order a weeks worth of tool box parts, we would order 10,000 sets of parts (2,000/day). When this material comes in, we add this 10,000 units to the inventory on hand. This would create a vertical line 10,000 units high from our on-hand inventory of the day.

4. *Normal usage* is a trend line indicating our balance on hand at the end of each day. In our tool box plant, we would be using up the parts at the rate of 2,000 sets per day.

5. *Minimum usage* rate is the slowest rate at which we use up parts. Normally, this would be only slightly less than normal usage, otherwise we wouldn't hit our 2,000 units per day goal. If we fall behind schedule, we would probably work Saturday to catch up (and use up the inventory).

6. *Maximum usage* rate is the fastest we would use parts. Again, if this was much faster than our plan we would be building up inventory of finished goods in our warehouse and an adjustment to our schedule would have to be made otherwise we would run out of parts. We want to carry a little extra stock so we will not run out. This is safety stock item 8 of this list.

7. *Normal distribution* between minimum usage rate and maximum usage rate around the normal usage rate is like any other normal distribution curve. This would indicate that our usage rate is faster than normal about half the time and slower than normal about half the time, but not by very much. To keep from running out and putting our plant down, we consider the maximum usage rate in determining our safety stock.

8. *Safety stock* is that extra inventory we carry so we will not run out of inventory, or run out only once in 100 order periods (or 1 percent outages). The distribution curve will tell us how big this safety stock needs to be in order to satisfy any level of service we choose.

9. *Reorder points* is that inventory level (in units on hand) where we need to reorder material to prevent a stock outage. While the order is being processed and shipped into our plant, we continue using inventory (depleting our stock). The reorder point is calculated by using the usage rate and the reorder time.

10. *Reorder time* is that time (in days) between the ordering of new material and the receipt of that material in our stores. If it takes 10 days to create a requisition, type a purchase order, mail the order to the supplying company, they'll fill the order, ship it to us, we receive it and put it in our store room, then we need 10 days of material on hand at time of reorder. In our tool box example of 2,000 per day and a safety stock of 1,000 units, our reorder point will be 21,000 units (2,000 × 10 days + 1,000 units). When our inventory drops below 21,000 units, we will reorder another quantity. The order quantity would be calculated by using a formula to minimize total cost, but that is subject for a production inventory control class.

11. *Stepped usage* is more realistic. As production needs parts, they request a days supply at a time. The inventory level drops all at once by a days supply; not by one unit at a time.

The inventory curve explains why and how we can provide room for only 50 percent of the required inventory. Look at the inventory curve (see Figure 7-8). How much inventory do you have on the day a new order arrived? How much inventory do you have on the day before the inventory came in? The answers are maximum or minimum. How much inventory do we have on the average? Answer, 50 percent. Now if we assign a spot in our store room for the maximum amount of inventory, how full will our store room be? On the average, only 50 percent full or one-half full. This is not good cube utilization. To get better use of our building cube, we allow room for only approximately 50 percent. Therefore, we cannot assign a part to any one location because there will not be enough room when the new supply arrives.

Provide Immediate Access to Everything

Our second design criteria for stores layout deals with *random locations*. Put anything anywhere, but keep track of where you put it. To clarify, pallets go in pallet racks, not on shelves, and steel storage is in another area. But within the racks, we can put anything anywhere.

A *location system* is needed to keep track of what you put where. A simple locator system would letter each aisle. Number each pallet location such as in Figure 7-9.

Figure 7-9 Storeroom Layout—Location System

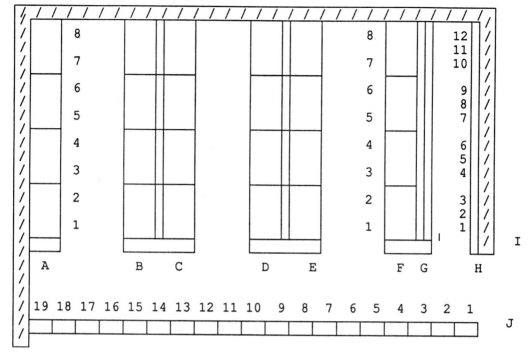

Rows A, B, C, D, E, and F are pallet racks (see Figure 7-10).
Rows G, H, I, and J are shelves (see Figure 7-11).

With four racks per row, two pallets per tier, and five tiers high, each row would hold 40 pallets. Row C, pallet six, "b" level, would be six pallets down row C and the second pallet up. Vertically, "a" would be the floor, "b" the second high and "e" the top tier. "e" would always be the top and "a" would always be the bottom.

Figure 7-11 is a six-tier shelf with each shelf measuring 3' wide, 1' deep and 1' high. Rows G, H, I, and J are shelves. Row I is a row of shelves on the end of racks.

Each location in our store room now has a location code. Our storekeeper is asked to put a pallet load of 1,500 part number 1750-1220 parts away. Our driver drives to the first open spot and deposits the pallet. Then the storekeeprs make out a location ticket like that in Figure 7-12. Two copies are needed: one copy would be attached to the pallet and one copy would be kept at the stores control desk in part number order.

Production now needs some of part number 1750-1220. The request comes to the inventory control desk. The storekeeper would look up part number 1750-1220 in the card file, find the pallet with the closest quantity to that requested or the

Figure 7-10a Industrial Pallet Rack

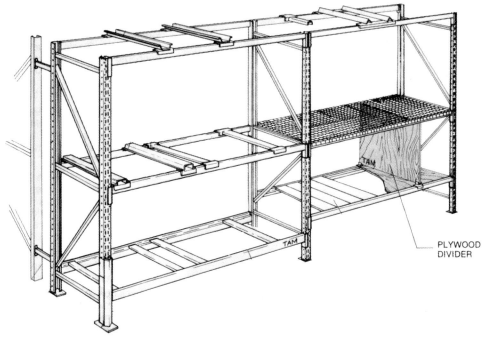

PLYWOOD
DIVIDER

Courtesy of Triple-A Manufacturing.

Figure 7-10b Storage Rack

Courtesy of White Storage & Retrieval System, Inc.

Figure 7-11 Industrial Shelves

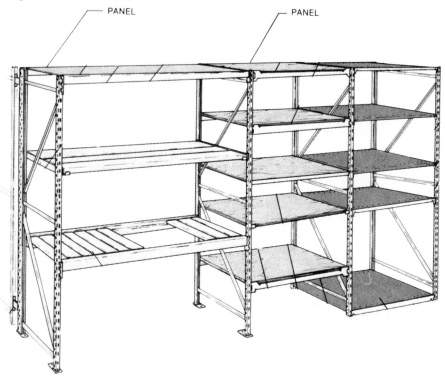

Courtesy of Triple-A Manufacturing.

oldest ticket and go to that location to retrieve the goods. The ticket can be pulled and sent to data processing to reduce the inventory. The inventory control department had previously added this inventory from a receiving report.

Storage facilities requirement spreadsheet. Every part must be measured for cubic size, multiplied by the number of parts to be stored, and converted to cubic feet (see Figure 7-14). The procedure for calculating store room size starts with an analysis of storage space needs as follows:

1. List all the raw materials and buyout parts. This will be column 1 (part number) and column 2 (part name).
2. After each part, list the length, width, height, and cubic inches of each part. This is columns 3, 4, 5, and 6.
3. Column 7 will be the quantity stated in your inventory policy divided by two (leaving room for only half the inventory).
4. Column 8 is cubic feet required. This is a result of multiplying column 7 by column 6 and dividing by 1,728 (cubic inches in a cubic foot).

Figure 7-12 Location Ticket

BLANK TICKET

PART #_____

QUANTITY_____

DATE_____

LOCATION_____

FILLED OUT

PART #_____1750-1220_____

QUANTITY___1500_____

DATE_____12/3/XX_____

LOCATION__B1C_____

5. Columns 9, 10, and 11 are the number of storage units required for each part. Column 9 would be shelf storage. Cubic footages under 10 cubic feet would be placed on shelves. Shelves are 3 cubic feet each ($1' \times 1' \times 3'$). Column 10 would be for pallets. Storage space requirements over 10 feet, up to 192 cubic feet, would be placed on pallets in the pallet racks (a pallet is $4' \times 4' \times 4'$ high or 64 cubic feet per unit load). Some items could be placed on the floor and stacked three pallets high and three pallets deep (see Figure 7-13).

The results of the storage facilities requirements spreadsheet is the number of shelves, pallet racks, and bulk storage areas needed. In Figure 7-14, 1,200 shelves, 1,000 pallet spaces, and 20 bulk storage areas are needed. The next step is to determine how many shelves to buy and how many pallet racks to set up. A shelving unit was pictured earlier in Figure 7-11. How many of these shelving units are required? (1,200 shelves divided by 6 shelves per unit equals 200 shelving units). The same thinking is used for pallet racks. Figure 7-10 showed 10 pallets per pallet rack. We need to store 1,000 pallets, so 100 pallet racks are needed. A storeroom layout is very close now. We know we need:

200 shelving units

100 pallet racks

20 bulk storage units

How will you lay this out?

Figure 7-15 shows how the storage space requirements for the tool box plant was developed. Figure 7-16 is the resulting stores layout in the tool box plant.

Aisle feet. The concept of aisle feet is very useful. *Aisle feet* will help determine the space needs. Visualize one shelf. Use Figure 7-11 if needed. One shelving unit is 3' wide. We need to place this open 3' on the aisle, therefore, one shelf has a need for 3 aisle feet. We need 200 shelves with 3 aisle feet each, so 600 aisle feet will be required. Another way of thinking about this is if we assemble 200 shelving units and placed them side by side. They would stretch out to be 600' long. A 600' row is too long, but how about (2) 300' rows or (10) 60' rows. There is almost unlimited flexibility of layout. Aisles for serving shelves can be much smaller than aisles serving pallet racks, so use 4' wide aisle until we discuss aisles in a later chapter.

How many aisle feet of pallet racks do we need? (100 pallet racks × 9' wide each equals 900 aisle feet). Again, (6) 150' rows or (15) 60' rows could be used. Floor storage units are 4' wide in the example. Twenty floor storage units are needed (4' × 20 = 80 aisle feet).

Figure 7-13a Floor Storage—Three Pallets High × Three Pallets Deep (4' × 12' × 12' = 588 cubic feet)

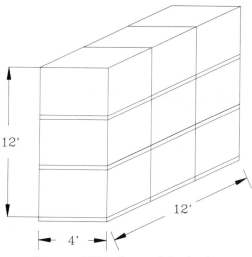

12'

12'

4'

Courtesy of White Storage & Retrieval System, Inc.

Figure 7-13b Floor Storage

Figure 7-14 Storage Facilities Requirement Spreadsheet

1	2	3	4	5	6	7	8	9	10	11
	Part							Shelf	Pallet	Floor
Part #	Name	L	× W	× H	= in.³	Q/2	ft.³	3 ft.³	64 ft.³	576 ft.³
1	Bracket	18	$\frac{1}{2}$	$\frac{1}{2}$		10,000				
2	Body	12	6	2		5,000				
3	Washer	$\frac{1}{2}$	DIA	$\frac{1}{8}$		20,000				
4	Nut	$\frac{1}{2}$	DIA	$\frac{1}{4}$		20,000				
5	Bolt	$\frac{1}{4}$	DIA	2		20,000				
6	Lid	12	6	1		5,000				
7	Hinge	6	1	1		10,000				
8	Handle	7	$\frac{1}{2}$	3		10,000				
9	Rivet	$\frac{1}{4}$	DIA	$\frac{3}{8}$		100,000				
10	Carton	24	16	$\frac{1}{4}$		10,000				
	etc. 990									
	other									
	parts									
1,000	Booklet	$8\frac{1}{2}$	11	.020		20,000				
	Total Storage Units							1,200	1,000	20

Fork trucks are needed to service pallet racks and floor storage areas and 8′ wide aisles are required for our equipment. At this time, all the information is available for you to lay out the storeroom. That information shows:

• 600 aisle feet of shelving;
• 900 aisle feet of racking;
• 80 aisle feet of floor storage; and
• 4′ and 8′ aisles.

Step 1: Start with a wall, placing the floor storage against the wall (see Figure 7-17).

Step 2: Place 900 aisle feet of pallet rack with 8′ service aisles. (Remember, racks are in multiples of 9′. 81′ is 9 sections of 9′ lengths. 100′ rows are not possible because 100 is not divisible by 9. 99′ or 108′ are possible.).

Step 3: Place 600 aisle feet of shelving serviced by 4′ aisles. Notice in Figure 7-17 that I used 124′ of shelving to create a wall between production and stores. This was designed to create a security system in which all movement is through a controlled door. Also note that I used 9′ of shelving as end caps to the rows of pallet racks. This is just good use of the aisles. Using both sides of an aisle is much more efficient.

Providing immediate access to everything was the second goal of a stores department. At receiving, everything was separated and checked in. The stores must maintain this separation and provide a separate location for different parts. The purpose of this goal is improved efficiency. When something is needed, the storekeeper should not need to stop, separate parts, then move them to production. That would take too much time.

Figure 7-15 Storage Space Requirement—22-day Supply, 2,000 Tool Boxes/Day

Part #/Box #	Part Name	L	× W	× H	= in.³	Q/2	ft.³ Needed	Shelf 1'×1'×3'	Pallet 4'×4'×4'	Floor 10'×3.5'×3' High
1	Handle	6	1	1	= 6	22,000	76			
2	Handle Clip	1	1	$1\frac{1}{2}$	= .5	44,000	12.7	5	2	
4	Rivet	6	6	6	(10,000)	88,000	1.1	1		
2	Catch	$\frac{3}{4}$	$\frac{1}{2}$	$\frac{1}{4}$	= .094	44,000	2.3	1		
2	Strike	1	$\frac{3}{4}$	$\frac{1}{4}$	= .188	44,000	5	2		
8	Rivet	6	6	6	(10,000)	176,000	2.2	1		
2	Hinge	6	$\frac{1}{2}$	$\frac{1}{8}$	= .375	44,000	9.5	4		
4	Rivet	6	6	6	(10,000)	88,000	1.1	1		
1	Name Tag	3	1	.02	= .06	22,000	.7	1		
1	Packing List	$8\frac{1}{2}$	$5\frac{1}{2}$	.005	= .234	22,000	3	1		
1	Booklet	$8\frac{1}{2}$	$5\frac{1}{2}$	.05	= 2.34	22,000	30		1	
1	Carton	36	24	$\frac{1}{4}$	= 216	22,000	2750		43	
1	Plastic Bag	10	6	.03	= 1.8	22,000	23		1	
#5	Steel*	5 lbs.	22 gal.			110,000	220			2
										2
				Total				17	45	

Our tool box warehouse needs only:

3 shelving units	= 9 aisle feet
5 pallet racks	= 45 aisle feet
2 floor storage area	= 20 aisle feet

A very simple layout, but steel will be received and stored in different areas.

* Steel weighs 500 pounds/ft.³ and comes in 42″ × 120″ × 18″ high.

123

Figure 7-16 Tool Box Stores Layout

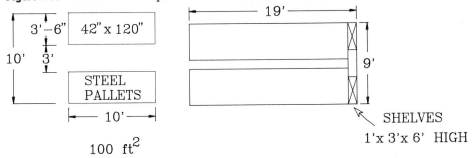

100 ft^2

SHELVES
1' x 3'x 6' HIGH

Provide Safekeeping

As seen earlier, our inventory is valuable. Good storage will provide safekeeping of this valuable asset. Having proper storage equipment like racks, shelves, and trucks will protect our product. Good containers can prevent dust and grime. The other part of safekeeping is preventing the unauthorized removal of inventory. Even the best intentioned supervisor can create inventory outages if they remove inventory without adjusting the inventory records. A security checkpoint and restrictions to entry are important parts of a storeroom design.

Figure 7-17 Storage Space Design

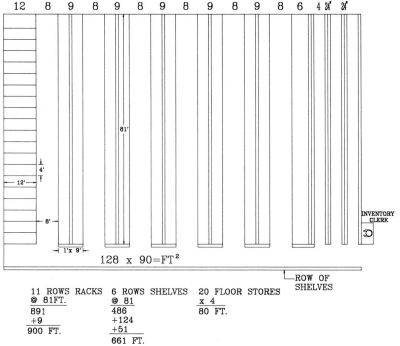

128 x 90=FT^2

ROW OF SHELVES

11 ROWS RACKS @ 81FT.	6 ROWS SHELVES @ 81	20 FLOOR STORES x 4
891	486	80 FT.
+9	+124	
900 FT.	+51	
	661 FT.	

Flat steel stock normally comes into the plant on 42″ × 120″ pallets. Tubing and bar sticks come in 12′ lengths. Special racks and special floor storage areas are needed for this material. Also, special material handling equipment will be needed. We will discuss material handling in Chapters 9 and 10.

WAREHOUSING

Warehousing is the storage of finished products. As in the storeroom, the area requirement will depend upon management policy. Seasonality could require us to stockpile our finished product for months in order to meet market demands. Sometimes outside warehousing is leased to carry the overload. No one would expect manufacturing to produce all the charcoal grills one month before the spring selling season. They have to be stored somewhere. Management must tell us how many units or how many day supply to allow space for.

A warehouse can be a department or an entire building. Our primary discussion will be about the department, but every technologist must know the important differences between these two warehouses. The *warehouse building* is where our company (which could have many manufacturing plants) sends its finished product. Our company may have many outside warehouses as well. Many manufacturing plants, sending their product to many warehouses in order to service our company's customers, is a function called *distribution*. The distribution system of a company tries to/minimize the cost of moving its product to their customers while maintaining superior customer service. A warehouse building will have a receiving department, a stores department, a warehouse department, a shipping department, and an office. The warehouse department in a warehouse building will have the same purpose as the warehouse department in a manufacturing plant.

The *warehouse department* (called just warehouse from now on) has the primary purpose of safekeeping our company's finished product. The stores department kept our raw materials and supplies, but the warehouse keeps our finished goods. After assembly and packout, our finished product is moved to the warehouse where it is kept until it is shipped to the customers.

Warehouse Design Criteria

Warehousing is the storage, order filling, and the preparation for shipping of our product. Order filling is the most labor intense portion of the job and affects our layout the most. Two design criteria are important to a warehouse layout:

1. fixed locations; and
2. small amount of everything.

No layout will ever be a single product layout. For example, a swingset manufacturing company made two basic types of swingsets called ''Big T'' and ''A'' Frame. Within each of these two groups, 50 different swingsets were sold.

The first warehouse design criteria (fixed locations) means every product must be assigned a fixed location so that the warehouse person can find that product quickly. Placing products in part number order is the simplest way, but not the most efficient. To increase productivity, the most popular items should be in the most convenient location.

The second design criteria is a direct result of the first criteria. By keeping only a small amount of everything in the fixed location, the order picker can pass all the product in fewer feet of travel. Just think if we kept only one pallet of every tool in the warehouse, 4′ times 8,000 items would require a trip of 32,000′ to pick one order. That's 6 miles! Let's be smarter and place these tools on 3′ wide shelves that stand 7 high. Now we would have to pass only 1,000 shelves, 3′ wide, or 3,000′. If we placed the shelves across the aisle from each other, only a 1,500′ trip would be required.

To further reduce the travel distance required to pick an order, an analysis of inventory can identify the most popular and profitable items and place these items in more convenient locations. This analysis is called *ABC inventory analysis*.

Figure 7-18 shows a simple warehouse. The top drawing shows a standard layout where the average part is in the middle of the warehouse. This would require a movement of 60′ from the middle of shipping to the middle of the warehouse in order to pick up a typical product to be shipped (120′ round trip). An ABC analysis (the bottom layout in Figure 7-18), would place the most important inventory (the A items) closest to shipping (20′ away) and the least important parts in the back of the warehouse (90′ away). The average distance to pick a product now becomes 28′ or 56′ round trip—a 50 percent savings. This was calculated as follows:

- A items account for 80 percent of the sales dollar and only 20 percent of the part number.
- B items account for 15 percent of the sales dollar and 40 percent of the part number.
- C items account for only 5 percent of the sales dollars but 40 percent of the part number.

A items = 20′ @ 80% = 16.0′
B items = 50′ @ 15% = 7.5′
C items = 90′ @ 5% = 4.5′

Total distance for
  average part   = 28.0′ (56′ round trip)

Functions of a Warehouse

The three basic functions of a warehouse are:

1. Safekeeping of finished product.
2. Maintaining some stock of every product sold by our company.
3. Preparing customer orders for shipment.

Figure 7-18 ABC Layout Cost Savings

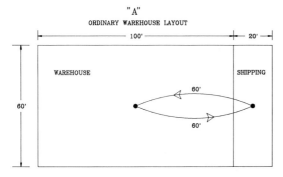

"A"

ORDINARY WAREHOUSE LAYOUT

The average trip in the ordinary warehouse would be from
the middle of the warehouse to the middle of shipping
(60 feet each way) 120 ft.
The average distance of an ABC layout is 28 feet one way
and 56 feet round trip

ABC LAYOUT

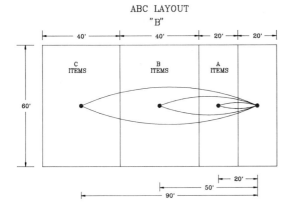

"B"

The safekeeping of finished product must consider pilferage as well as damage due to material handling and storage facilities. Containers, shelves, racks, fences, gates, control desks, and inventory control systems are all a part of this safekeeping requirement and responsibility of warehousing.

Picking orders as requested by the customers is a function of warehousing that affects the layout of that warehouse the most. The efficiency of the warehouse will be determined by the layout. An example of a warehousing job may be a book publisher's warehouse. The warehouse may have 4,000 different titles. Each title is called a *stockkeeping unit* (SKU), so the book warehouse would have 4,000 SKUs. The big question is, "How do we layout these 4,000 SKUs in order to be able to pick customer's orders efficiently?" A simple-minded solution would be to place the books on pallets, and place these 4,000 pallets next to each other. Four thousand, 4' by 4' pallets lined up would be 16,000' long even if we used both sides of the aisle. An 8,000' aisle would be needed. Our order pickers would have to walk 8,000' down the aisle and 8,000' back the other side to pass every book title. Over 3 miles of walking per order is not a good use of our people, so the first design criteria or warehouse layout is:

Keep a small amount of everything in a small fixed location.

A *small amount* may be defined as a one to five days supply. This inventory could be placed on shelving or, better yet, in flow racks (see Figure 7-23).

One 6' wide flow rack, six high would have 36 different SKUs (titles) in one 6' area, so 112 of these racks could store 4,000 SKUs.

$$\frac{4,000 \text{ SKUs}}{36 \text{ SKUs/rack}} = 111.11 \text{ or } 112 \text{ racks}$$

112 rack × 6 ft./rack = 672 ft. of rack.

Figure 7-19 shows a layout which greatly reduces the 16,000' of travel required in our preview to only 678' of travel per order. This is still too much, but look how much we have improved already.

The next improvement in ABC analysis is called the *80/20 rule* or *Pradeo analysis*. All of these mean the same thing. The 80/20 rule simply states that 80 percent of our sales (measured in dollars) come from 20 percent of our product (book titles, in our example). To maximize our efficiency, we want to identify those products that are accounting for most of our sales.

This 80/20 rule divides inventory into three categories.

Class	Percentage of $	Percentage of Parts
A	80	20
B	15	40
C	5	40

The "A" category of inventory is exactly like the 80/20 talked about already, but the less popular 80 percent of the products are further divided into "B" items and "C" items.

Now, the distance of travel to pick the average order is:

A items = 80% × 138' =	110.4'
B items = 15% × 100' =	15.0'
C items = 5% × 270' =	13.5'
Total distance of travel =	138.9'

A comparison of methods shows that we have reduced the walking from 16,000' using pallets alone to 678' using flow racks only, to 138.9' using ABC analysis, flow racks, and pallets. Let's work smarter, not harder.

Figure 7-19 Flow Racks—Layout For Book Publisher

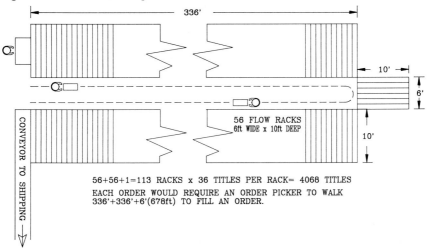

56+56+1=113 RACKS x 36 TITLES PER RACK= 4068 TITLES
EACH ORDER WOULD REQUIRE AN ORDER PICKER TO WALK
336'+336'+6'(678ft) TO FILL AN ORDER.

Procedure for Sales Analysis of ABC Inventory

To conduct a sales analysis using ABC inventory:

1. List all products with their unit price and average monthly demand (sales).
2. Multiply the price times the average monthly demand.
3. List the product in order of most monthly sales dollars first and the least monthly sales dollars last.
4. Add up all the monthly sales (total sales).
5. Run a cumulative column after the total monthly sales, add all the previous totals to each line. For example:

Part #	$/Unit	Mo. Sales	Total $	Cum. $	Percentage of Total
1650	34.50	2,000	69,000	69,000	28
1725	49.90	1,000	49,900	118,900	49
1400	45.00	1,000	45,000	163,900	67
0390	20.50	2,000	41,000	204,900	84
1450	39.00	1,000	39,000	243,900	100
			243,900		

6. The percent of total column is the cumulative dollars divided by the total dollars. In a real example we would see that only 20 percent of the part numbers accounted for 80 percent (the cumulative percent column) of the sales dollars.

Our book example with 4,000 books would have the following breakdown:

# Book Titles	$	Percentage of $	Percentage of Book Titles
800	8,000,000	80	20
3,200	2,000,000	20	80

To lay this out, we would place these 800 books close to the shipping department (see Figure 7-20).

Placing product in part number order is the simplest way of laying out a warehouse, so when a customer order comes into the warehouse printed in part number order, the picker goes to the first part number, then to the second, etc. The products are easy to find because its in part number order. The problem with this organization of the warehouse is that slow moving parts are right next to fast moving parts. To fix this problem, every warehouse location can be numbered and a product number can be assigned to any location. In this system, the most popular items can be placed in the most convenient locations. As the ABC analysis is made, the "A" items are given convenient locations, while the "C" items are located in the back of the warehouse because we have to go to the "C" area only 5 percent of the time; the "C" area is 40 percent of our warehouse. When the

Figure 7-20 ABC Book Warehouse Layout

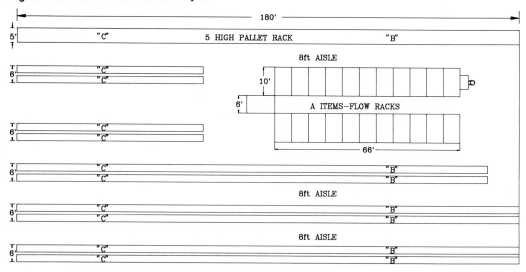

"A" ITEMS= 800 BOOKS IN FLOW RACKS (36 BOOKS PER RACK x 23 RACKS)
"B" ITEMS= 1600 BOOKS ON PALLET RACKS 5 HIGH= 640ft RACKS
"C" ITEMS= 1600 BOOKS ON PALLET RACKS 5 HIGH= 640ft RACKS

customer's order comes into our warehouse, the product is in picking location order: The picker is told to go to location number 0529 and pick up six part number 1650-1900s, then go to location number 0533 and pick up 12 part number 1700-1550s, etc.

ABC Inventory Layout of a Hand Tool Manufacturing Company's Warehouse

This company provides 8,000 different tools to their customers. The company markets three different brand names. The only difference in some tools are the names. Their old layout was divided into three areas (one for each brand name) and within each area the tools were stocked on 3' wide shelves 1-1/2' deep and 1' high. The shelves were six high and the tools were placed on these shelves in part number order. Part number 1 was the first tool on the shelf, and many aisles later, part number 9,999 was the last tool in the warehouse. Figure 7-21 is a layout of one of three brands of tools. An order picker would pick up a customer's order from the warehouse supervisor's desk, and walk 3,000' through the entire section of shelving to pick an order. Figure 7-22 is an improved ABC inventory layout for the same section (brand name of tools). In this layout the "A" items were located on the main aisles, the "B" items were located on the side aisles but close to the "A," and the "C" items were located behind the "B" items. Locations were numbered and the customer orders come out of data processing in location number order.

The proposed layout (see Figure 7.22) requires the picker to walk only 5.4' into each side aisle. These aisles × 5.4' = 227' + 330' up the main aisle and back equals 557' compared to 3,000 feet of travel in the present layout.

"A" Items = 80% walk 3' = 2.4'
"B" Items = 15% walk 12' = 1.8'
"C" Items = 5% walk 24' = 1.2'

    Total Average Distance = 5.4'

Our pickers will need to walk only 18.6 percent of the distance they used to walk (less than 1/5 of the old distance). This will result in fewer pickers being needed. This also resulted in less area being required. A world class distribution center means that we can compete with the best warehouses in the world. We must work both smarter and harder to be the best.

A small amount of everything is the prime criteria for warehouse layout. A "small amount" could mean a one-day supply up to a one-week supply, but never everything we have of that part number. If we don't leave room for everything in the warehouse, where does the excess inventory go? We may have a 30-day supply of one part number and our warehouse is designed to hold only a one-day supply. This excess inventory would be called *bulk stock* or *backup stock* and could be kept in the raw material storeroom. Remember that stores used random

Figure 7-21 Hand Tool Company Warehouse—Present Layout (Part # Order—8,000 Items on 1,000 3′ × 1.6′ × 7′ High Shelves 3,000 Aisle Feet Are Needed Or 40(75′) Rows of Shelves)

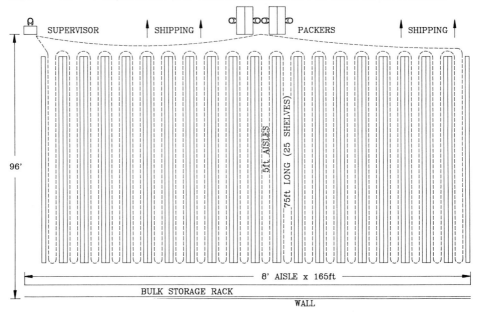

Figure 7-22 Hand Tool Company's Warehouse—ABC (Proposed) Layout (Same Layout As Figure 7-21, But Two Sections Of Shelves Removed to Create New Aisle and Place All "A" Items In The First 3′ Off the Aisle)

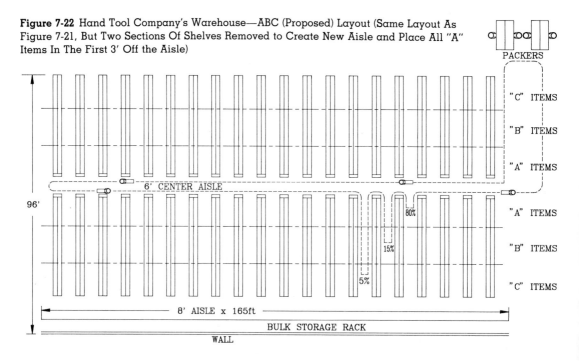

storage, so we can put it anywhere. Special sections of the warehouse can be set up as bulk stock or backup stock areas. These areas would be layed out and controlled just like storerooms.

Keeping the shelves full is the job and responsibility of a group of warehouse people other than the order pickers. These warehouse people move material from the bulk areas to the shelving picking area. Sometimes, these warehouse stockers pull very large orders directly from the bulk areas so as not to deplete the shelf stock.

Warehousing is judged by service level and pounds shipped per labor hour. Whatever can be done to improve these figures will be good for the company.

Warehouse Space Determination

The size of the finished/packaged product multiplied by the quantity manufactured each day times the number of days supply will equal the cubic footage of warehouse space required.

Example: Provide a warehouse to store a 30-day supply of tool boxes at the rate of 2,000 units per day.

$$\frac{18'' \times 8'' \times 8''}{1,728 \text{ ft.}} = .666 \text{ cubic feet each} \times 2,000 \times 30 \text{ days}$$

$$40,000 \text{ cubic feet} + \text{pallets}$$

See layout in Figure 7-23 for a pallet pattern and a typical pallet.

$$\frac{42'' \times 42'' \times 54''}{1,728 \text{ ft.}} = 55.125 \text{ cubic feet/pallet}$$

$$72 \text{ tool boxes per pallet}$$
$$2,000/\text{day} \times 30 \text{ days} = 60,000 \text{ tool boxes}$$

$$\frac{60,000}{72} = 833 \text{ pallets}$$

Figure 7-24 is a layout for the tool box plant's warehouse. Note that the pallets are 42″ × 48″ (a standard width) and that 8 pallets deep is only 28′. Pallets are stacked right next to each other (no room in between) because everything is the same and can be a solid stack. This is a very simple one product layout. More complicated layouts are just more of the same procedure. If we calculate the cubic space required for each of our products and total them, we will have the total storage space. Doubling this space will allow for aisles, and 50 percent aisles is more normal than our example. When you have only one product, you can store deep (8 pallets from the aisle). More normally, we can store only one of two pallets deep, thereby requiring much more aisle space.

Figure 7-23 Pallet Pattern—Tool Boxes (12 Per Tier, 72 Per Pallet)

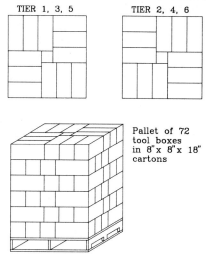

TIER 1, 3, 5

TIER 2, 4, 6

Pallet of 72
tool boxes
in 8"x 8"x 18"
cartons

Warehouse Equipment

Shelves which look very much like book shelves in a library are most common for picking areas (see Figure 7-11). Adjustable shelves allow for different sizes and quantities. Shelves are usually all the same size. For more room, most shelves must be used. Sometimes, three or four different parts can be stored on one shelf.

Figure 7-24 Warehouse Layout—Tool Box Plant

48'x 42' pallets stacked 3 high= 816 pallets

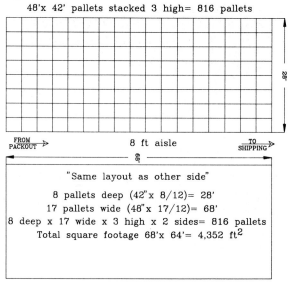

28'

FROM
PACKOUT →

8 ft aisle

TO
SHIPPING →

68'

"Same layout as other side"

8 pallets deep (42"x 8/12)= 28'
17 pallets wide (48"x 17/12)= 68'
8 deep x 17 wide x 3 high x 2 sides= 816 pallets
Total square footage 68'x 64'= 4,352 ft^2

Tool warehouses use very heavy shelving which measure 3' wide, 1-1/2' deep and 1' high with an average of seven shelves high. Each shelf will hold 4-1/2 cubic feet of parts. A one-week supply is warehoused on the shelves. The overstock is kept in the storeroom.

A *mezzanine*, a form of balcony, can be built over a shelving area for additional shelves. Slower moving inventory can be stored upstairs so as to make good use of an otherwise poorly used space.

Two-wheeled hand carts are often used to stock shelves. Boxes of material may be brought to the warehouse department by *fork truck,* but the aisles are not big enough to allow fork trucks, so hand carts are used. The cartons are moved to the shelves. Heavy (over 25 pounds) cartons are unpacked and placed on the shelves by hand.

Picking carts are four-wheeled shelf carts which are pushed around the shelves to pick customer's orders. The carts are unloaded as the packer fills cartons to ship to the customer.

Racks are used for larger products. The spacing between shelves can be as large or small as needed, but 2' or 3' is common. Two or three high is all that can be stacked because of picking height restrictions. Tool boxes are stored on racks in our tool company warehouse.

Flow racks allow for many parts to be warehoused in a small location. In a drug warehouse, 2,000 of the most popular drugs were warehoused on a 50' aisle. Eighty percent of the sales dollars was shipped out of this very small part of the drug warehouse (see Figure 7-25).

Figure 7-25a Flow Rack

Courtesy of Hytrol Conveyor Co.

WAREHOUSING

Figure 7-25b Flow Rack Picking Area

Courtesy of S.I. Handling Systems, Inc.

Conclusion

Warehousing is an area where a little planning and creative thought can save a lot of space and improve efficiency. The main design criteria are:

1. Allow a small fixed location for everything.
2. Divide the inventory into ABC classifications.
3. Locate the "A" items closest to shipping in the most convenient location. Example of ABC Warehouse

	Percentage of $	Percentage of Part	Warehouse Days Supply
A	80%	20	2
B	15%	40	5
C	5%	40	10

4. Calculate the storage space required for each item in the warehouse and multiply the unit cubic foot by the number of days supply.
   Examples:

   1.5 cubic feet would be 1/3 of a shelf
   6.0 cubic feet would be 1 1/3 shelves

5. Calculate the total number of shelves.
6. Determine aisle size.
   a. One-way aisles should be 3 to 4' depending on size of material.
   b. Two-way aisles should be one foot wider than two pieces of material handling equipment. An 18″ picking cart would require a 4' aisle.
7. Layout the shelves and the aisles and determine the warehouse width and length.
8. Maximize the warehouse cubic space. Mezzanines and racks can best use the space over our heads.

MAINTENANCE AND TOOL ROOM

The maintenance and tool room function is to provide and maintain production tooling. These functions vary widely from one company to another. Tool rooms may not exist in some plants because all tools are purchased from outside sources. Some maintenance is also contracted to outsiders. For example, office cleaning is usually done by an outside firm.

Sizing the maintenance and tool room is dependent upon management's desire to do it in-house or to contract out all or part of these jobs. A tool room is made up of machines and an assembly area just like production. Once management determines what our plant will do, a machinery list is determined and each machine needs a workstation design. The tool room size is the sum total of all the equipment space requirements times 200 percent. The extra space is for everything except raw material and finished tool storage areas. These areas are calculated just like every other storage area.

Maintenance is service to the company's equipment. A mobile service cart could be our maintenance program, but more commonly a central maintenance area would include equipment, machine overhead areas, maintenance supply, and spare parts storage areas. Maintenance can account for 2 to 4 percent of the plant personnel. As an extreme example, maintenance accounted for 33 percent of a paper mill's employment. If we know the size of the plant (number of employees) and, from corporate experience (or industrial averages), that our company should have three maintenance people for every 100 production people, we could provide them with 400 square feet of space each. This would allow for everything except maintenance stores which is calculated like any other storeroom.

Our tool box plant used .13702 hrs./unit of the rate of 100 percent. History indicates that 85 percent performance is more realistic, so

$$\frac{.13702}{.85} = .16120 \text{ hrs. each}$$

.16120 × 2,000 box/day = 322.4 hours of production people. Each person works eight hours per day, so 41 people are needed. Three percent of 41 people equals 1.2 maintenance people required. So we will allow space for two maintenance people.

$$2 \times 400 \text{ square feet} = 800 \text{ square feet}$$

Add a 10′ × 10′ controlled storeroom for tools and supplies to most plant layouts. This 100 square foot storage area is just a minimum size area for controlling supplies. Our plant will buy its tools, so no tool room is needed. Our total square footage for maintenance will be 900 square feet.

UTILITIES

Heat, air conditioning, electrical panels, air compressors, and the like must be considered when determining space. These areas also must be kept separate from the normal traffic—electrical panels should be fenced off, heaters must be kept clean, air compressors require special construction because they are noisy. Once these facilities have been identified, they are sized and placed in an appropriate area of the plant. Many times, we can place these facilities out of the way (on the roof or in the trusses) so they don't interfere with material flow. But, remember whether in plain site or tucked out of the way, utilities must not be overlooked when determining plant space.

QUESTIONS

1. What are auxiliary services (support services)?
2. What does shipping and receiving have in common?
3. What are the advantages of a common receiving and shipping department?
4. What are the disadvantage?
5. Should we have only one receiving area?
6. What affect does the trucking industry have on our receiving and shipping docks?
7. What is LTL?

8. Why would we use common carriers?

9. What are the functions of a receiving department?

10. What is a Bates log?

11. What is a Julian calendar date?

12. What is an OS&D report?

13. What is a receiving report?

14. How many dock doors should we have?

15. What does arrival rate mean?

16. What outside areas are needed for receiving and shipping departments?

17. What is the visualization method of determining receiving department space requirements?

18. What are the functions of a shipping department?

19. Why do we weigh shipping containers?

20. What is a bill of lading?

21. What is a store?

22. What are the different types of stores?

23. What determines the stores size?

24. What is ABC classification?

25. What is inventory carrying cost?

26. What is JIT?

27. What are the goals of a stores department?

28. Review Figure 7-8 (the inventory curve) and identify:

    a) the order quality

    b) the usage rates (normal, maximum and minimum)

    c) safety stock

    d) reorder point

    e) reorder time

29. How can we get away with leaving room for only 50 percent of the inventory?

30. How does random location work?

31. What is an aisle foot?

32. Layout a storeroom with 18 bulk storage areas (4' on the aisle × 12' deep) + 800 aisle feet of pallet racks for 4' × 4' pallets + 400 ' of shelving (1' × 7' × 3' wide). Use 8' aisles for fork trucks and 4' aisles for shelves. Calculate the square feet.

33. What is a warehouse?

34. What are the two design criteria for a warehouse?

35. What are the two functions of a warehouse?

36. What is order picking?
37. How does ABC inventory analysis help us lay out the warehouse?
38. What is a pallet pattern?
39. What is a mezzanine?
40. How many maintenance people should a plant have?

CHAP. 7: AUXILIARY SERVICES—SPACE REQUIREMENT

Chapter 8

Employee Services—Space Requirements

Our employees have needs, and employee services describe the various needs. This chapter will discuss:

1. Parking lots
2. Employee entrance
3. Locker room
4. Toilets
5. Cafeteria
6. Recreation areas
7. Drinking fountains
8. Aisles
9. Medical services
10. Break areas and lounges

These services require quite a bit of space. Their locations will affect the efficiency and productivity of our employees and the quality of these services will affect the employee's relationships with the company's management. It is said that if you want to "see" management's attitude toward their employees, look in the bathrooms. If it is untidy or in disrepair, a poor attitude exists. A neat, clean area indicates a positive attitude.

The interface between the outside world and our plant is the driveways and parking lots. Our goal is to provide adequate space with a convenient location. Three parking lots may be needed. They could be broken down by usage as follows:

1. Manufacturing employee parking
2. Office employee parking
3. Visitor parking

Convenience is a very important consideration when determining the parking lot design, so the entrances to the plant will determine where the parking lots are located. Parking as close to the entrance as possible should be our goal, but remember not everyone can park in the same space. One thousand feet takes an average of four minutes to walk. That distance should be the furthest point for either employee or visitor to walk. Large plants could have trouble with this requirement due to space restrictions, but it still should remain a goal.

The size of the parking lot is directly proportional to the number of employees. If we were located out in the country and our employees drove to work, we might allow one parking place for every one and one half employees. If we were closer to town and property was costly, we might allow one parking space for every two employees. We must consider our plant location, the number of employees, and our managements attitude toward car pools and the like and then decide on the parking space/employee ratio.

Employees		Parking Spaces	Spaces/100 Employees
1.25	to	1	88
1.5	to	1	75
1.75	to	1	63
2.0	to	1	50

Office parking may be different from factory parking because we can incorporate the visitor parking spaces in this area. A ratio of one to one may be appropriate and we could assign the closest parking spaces to visitors parking.

Assigned parking is normally a bad idea, especially if employees who arrive early have to park further away from the employee entrance and pass these "big shots" parking places. Assigned parking indicates an air of superiority. This could lead to poor employee relations. The best assignment is first come, first served for the best spot. Another way of improving parking and employee relations is to have more than one entrance (i.e., a factory entrance and an office entrance. This way, the later arriving office people still get a chance at a good spot. The only two reasons for assigned parking spaces that I can think of are for company cars and car pool vehicles. The car that is used to run errands should be conveniently located close to the entrance to promote productivity. Car pooling promotes cost efficiency.

Once the number of parking lots and the number of parking spaces has been determined, there is almost an unlimited number of ways to arrange parking. The size and shape of the available space may be the deciding factor, but some general statistics follow:

Parking Stalls	Width	Length
Small cars	8	15
Medium cars	9	17.5
Large cars	10	20

Width of Driveways

Single lane—11'
Double lane—22'

1. As the angle of the parking spots increases, the width of the aisle needs decreased.
2. As the width of the parking spots increase, the width of the aisle needs decreased.
3. The wider the parking spots (to 10'), the less door damage in your lot.
4. Local building codes often dictate parking spot size.
5. Local building codes often dictate the number and location of handicap spots.
6. As a general rule of thumb, a parking lot will be 250 square feet per number of parking spots needed (see Figures 8-1 and 8-2).

Figure 8-1 Perpendicular Parking (Most Space Consumed)

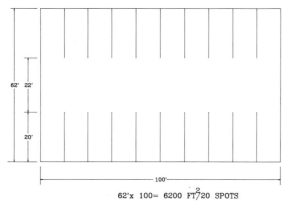

62' 22'

20'

100'

62'x 100= 6200 FT^2/20 SPOTS

Figure 8-2 Angular Parking Lot

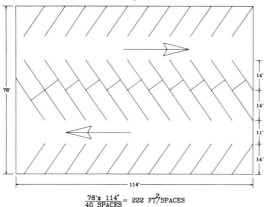

$$\frac{78' \times 114''}{40 \text{ SPACES}} = 222 \text{ FT}^2/\text{SPACES}$$

EMPLOYEE ENTRANCES

Where the employees enter the plant will have an affect on the placement of parking, the locker room, time card racks, restrooms, and cafeterias. The flow of people into a factory is from their cars into the plant via the employee entrance to their lockers and to the cafeteria to wait for the start of their shifts.

The employee entrance is where security, time cards, bulletin boards, and sometimes the personnel departments are located. Depending on management's attitude and corporate requirements, the employee entrance can vary from a simple doorway with a time card rack and time clock to a series of offices and gates through which to pass.

The size of the employee entrance must consider individual requirements. How many persons will be using this door at any given time? The door could measure from 3' to 6' with an aisle or walkway leading into the plant. Allow for traffic flow.

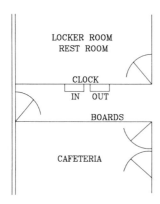

Figure 8-3 Simple Employee Entrance

CHAP. 8: EMPLOYEE SERVICES—SPACE REQUIREMENTS

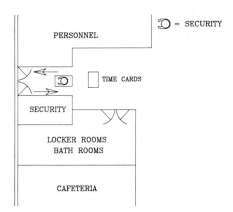

Figure 8-4 Employee Entrance With Security

Personnel offices and security offices will be sized at 200 square feet per office person. About one personnel person per 100 employees, and one security person per 300 employees is normal.

Our tool box plant employees entrance will be a 4' door and a 6' aisle like the layout in Figure 8-3. The time clock and time card racks will be mounted on a wall and on the other side of the aisle, the bulletin board will be mounted on a wall. Our employee entrance will be 6' × 15' or 90 square feet. Figure 8-4 shows a larger plant's employee entrance with security.

LOCKER ROOMS

Locker rooms give the employees space to change from their street clothes to their work clothes, and a place to keep their personal things while working. Their coats, lunches, street shoes, and so forth will be kept in their lockers. Locker rooms are very much like gym locker rooms. Showers, toilets, wash basins, lockers, and benches are all a part of a well-equipped locker room, but we will consider toilets and wash basins in the next section.

The size of a locker room is directly proportional to the number of employees. Figure 8-5 shows a typical small locker room. The top tier could be for day shift and the bottom tier for night shift. Staggered shift starting could reduce congestion in the locker room. For example:

7:00–11:00	Lunch	11:30–3:30
7:30–11:30	Lunch	12:00–4:00
8:00–12:00	Lunch	12:30–4:30
3:30–7:30	Lunch	8:00–12:00
4:00–8:00	Lunch	8:30–12:30
4:30–8:30	Lunch	9:00–1:00

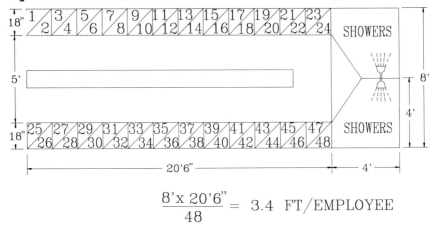

Figure 8-5 Locker Room

$$\frac{8' \times 20'6''}{48} = 3.4 \ \text{FT/EMPLOYEE}$$

In a plant of 48 people, only 8 people would be in the locker room at one time. If this was a plant of 96 people (48 men and 48 women) two identical locker rooms would be provided.

The size of a locker room can be initially sized by multiplying the number of employees times 4 square feet per employee.

TOILETS AND RESTROOMS

Our discussion of toilets will include urinals and wash basins. These facilities are usually called restrooms, but the term "restroom" is not properly descriptive. How many restrooms may be our first question. Restrooms should be no further than 500' away from any employee. These 500' take two minutes to walk, so such a distance can be very costly. At a minimum, we would have a men's and women's restroom in the office and in the factory.

How many toilets will depend on how many employees work on the major shift. The local building code may dictate how many toilets are necessary. The number of wash basins is equal to the number of toilets. See Figure 8-6 for a partial example of a local building code.

Figure 8-6 Sample Building Codes For Toilet Accommodations

Sec. 31-37 Toilet Accommodations in Manufacturing Etc.:

". . . Provide adequate toilet accommodations, so arranged as to secure reasonable privacy for both sexes . . . inside such establishment when . . . practicable . . . adequate fixtures . . . good repair . . . clean and sanitary condition, adequately ventilated with windows or suitable ventilators opening to the outside . . . provide convenient means for artificial lighting . . . clearly marked . . . to indicate the sex for which . . . intended for use . . . and . . . partition . . . to be solidly constructed from the floor to the ceiling. . . ."

Sec. 177-4-5 Toilet Facilities:

a. Water-closets required and sex designation:

"Separate water-closet . . . for each sex . . . clearly marked. . . ."

Figure 8-6 (cont'd) Sampling Building Codes For Toilet Accommodations

    b. Number:

        "Water or toilet closets shall be provided for each sex at the rate of one closet to twenty persons or fraction thereof, up to 100, and thereafter at the rate of one closet for every twenty-five persons."

    c. Location:

        "Such closets and urinals must be readily accessible to the persons for whose use they are designed. In no case may a closet be located more than three hundred feet distance from the regular place of work of the persons for whose use it was designed. . . ."

Sec. 177-4-6 Privacy:

    b. New installations:

        1. "Every water-closet compartment . . . shall be located in a toilet room, or shall be built with a vestibule and door to screen the interior from view."

        2. "The door of every toilet-room and of every water-closet compartment, which is not located in a toilet-room, shall be fitted with an effective self-closing device to keep it closed."

        3. ". . . male and female . . . adjoining compartments (must have) solid plaster or metal covered partitions . . . extending from floor to ceiling."

Sec. 177-4-7 Construction:

    b. New installations:

        1. "The floor . . . and the side walls to a height of not less than six inches . . . shall be marble, portland cement, tile, glazed brick, or other approved waterproof material."

Sec. 177-4-12 Urinals:

    a. "In establishments or departments employing ten or more male employees, one urinal shall be installed for every forty males or fractional part thereof up to two hundred and thereafter, on additional urinal for every sixty males or fractional part thereof. Two feet of an approved trough urinal shall be equivalent to one individual urinal. . . . Urinals may substitute for up to 50% of men's toilets."

Sec. 177-4-4 Washing Facilities:

    a. "All water supplied by any establishment for washing purposes shall be potable (pure, clean water suitable for drinking or for washing purposes)."

    b. "Every establishment should furnish for each sex at least one standard wash basin or its equivalent for every 20 such employees or, fractional part thereof, up to 100. Beyond 100 the ratio may be one basin or equivalent to each 25 employees of either sex, or fractional part thereof."

    c. "If washing sinks or troughs are furnished, each 2-1/2 feet of trough or sink equipped with a hot water and a cold water faucet or a single faucet carrying tempered water may be counted equal to one basin. Where washing fountains are furnished, 2-1/2 feet of the circumference of such fountains shall be equivalent to one wash basin.

Sec. 31-41 Order to Remove Excessive Dust:

"Each employer whose business requires the operation or use of any emery, tripoli, rouge, corundum, stone, carborundum or other abrasive, polishing or buffing wheel in the manufacture of articles of metal or iridium or whose business includes any process which generates an excessive amount of dust, shall install and maintain . . . such devices . . . necessary . . . to remove from the atmosphere any dust created by such process. . . ."

Sec. 31-45 Emergency Kits Required by Factories:

"Each . . . firm . . . employing persons to work . . . with dangerous machines . . . , except those maintaining equipped first-aid-to-the-injured rooms, shall . . . (place) where such machinery is operated . . . an emergency kit for use in case of accidents . . . such (kits) shall be kept in a dustproof case or cabinet within easy access. . . ."

Par. 716.3 Locks and fastenings on required exit doors shall be readily opened from the inner side without the use of keys. Draw bolts, hooks, and other similar devices shall be prohibited on all required exit doors.

728.2 All exit signs shall be generally located at doors or exit ways so as to be readily visible and not subject to obliterations by smoke. They shall be illuminated at all times when the building is occupied from an independently controlled electric circuit or other source of power.

Figure 8-7 Men's Restroom

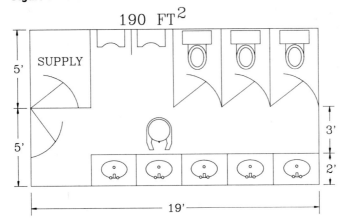

The size of a restroom is 15 square feet per toilet, wash basin and entry way, and 9 square feet for urinals.

If our plant has 50 male employees and 50 female employees, then our two restrooms would look like this:

	50 Men	50 Women
Toilets	2 @ 15 = 30	3 @ 15 = 45
Wash Basins	3 @ 15 = 45	3 @ 15 = 45
Urinals	1 @ 9 = 9	
Reclining area		1 @ 15 = 15
Door	1 @ 15 = 15	1 @ 15 = 15
Total	99	120
×150%	149	180

See Figure 8-7 for a diagram of this men's restroom.

LUNCH ROOMS

There are five types of eating facilities:

1. Cafeterias with serving lines
2. Vending machines
3. Mobile vendors
4. Dining rooms (executive)
5. Off-site diners (lunch counters)

A cafeteria feeds a lot of people in a short time. Schools, military installations, and family picnics use this type of service. Many people line up at a serving line and are given food as they pass different stations. One serving line can service 9 employees per minute (about 7 seconds each). One line is 30' long and 10' wide. Employees would not spend more than 10 minutes in line, so 90 people could be served every 10 minutes. If lunch periods were staggered, 540 people per hour could be served. Cafeterias are generally used in big plants.

Vending machines can serve very complete meals. A vending machine with a microwave oven for special foods can provide employees with many meal choices. Vending machines are generally used for small plant lunch rooms. The machines are lined up against a wall. Room for customer lines must be allowed in front of the machines. Vending machines earn money. Most companies use these profits for employee benefit programs, but these profits could be used to buy services from the vending company (like a custodian who would be available to service the machines in high use periods). These people can also be used to keep the lunch room in good order.

Mobile vendors are outside vendors who drive their specially built pickup trucks to our back door and honk their horns signifying the start of lunch periods or breaks. Only very small plants could use this service. Office buildings also use smaller mobile carts for serving coffee and donuts. They can go from office to office and even from floor to floor. The mobile vendors are affectionately called "Roach Coaches." Whatever they are called, they provide a useful service.

Executive dining rooms are very special and are used to entertain special customers, vendors, and stockholders. They usually provide a limited selection from a menu with the meal being cooked on site. Dining rooms cannot serve many people at a time and usually eating takes more time.

Off-site dining at local diners is attractive to many employees. It allows them to get away from the job. Private business people develop local clientele for lunch and create a comfortable environment in which to eat. But, most factory people are not given enough time to leave the plant for lunch. Companies also discourage people from leaving the plant at lunch because they lose control.

Lunch rooms should provide a comfortable, pleasing environment in which to recuperate from work and to eat lunch. Nice facilities show respect for employees and they improve the productivity of the work force by allowing them to regenerate their energy for the next work period. Comfort, attractiveness, speed of service, and convenient location are all important in the design of a lunch room. The location is analyzed in the activity relationship diagram studied in Chapter 5, but two more factors should be considered.

1. An outside wall would allow for an outside eating area. During nice weather, this facility would increase morale. An outside wall gives easy access to delivery of food and the removal of trash.
2. The employee flow typically calls for the employees to wash up before eating, and getting their lunches (brown bag) from their lockers, so restrooms and locker rooms should be close to the lunch room.

Figure 8-8 A Staggered Lunch Room

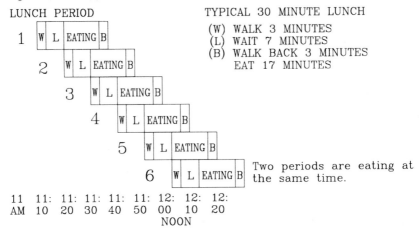

LUNCH PERIOD

TYPICAL 30 MINUTE LUNCH

(W) WALK 3 MINUTES
(L) WAIT 7 MINUTES
(B) WALK BACK 3 MINUTES
EAT 17 MINUTES

Two periods are eating at the same time.

The size of the lunch room will depend on

1. The number of employees;
2. The type of service provided; and
3. The facilities included.

Figure 8-9 shows the backup data for three different sized lunch rooms. As can be seen, they are very similar in size per person (10 square feet). Cafeteria space can be saved if the food is cooked off-site and carried into the plant at lunch time. Space can be saved by overlapping lunch periods. A lunch period starting every 10

Figure 8-9 Lunch Room Space Determination

Per Person or Unit	3 Periods 100 people	5 Periods 500 people	7 Periods 1,000 people
Cafeteria Serving Line (300 ft.)	—	300	600
Waiting Line (4 ft.²)	120	180	320
Vending Machines (20 ft.²)	100	—	—
Eating Area @ (15 ft.²)	495	1,995	3,000
Waste (1/2 ft.²)	50	250	500
Food Storage (1/2 to 1 ft.²)	50	500	1,000
Food Prep. (2 ft./meal)	—	Catered	2,000
Dish Washing (1/2 to 1 ft.²)	—	500	750
Total	815	3,725	8,170
Aisles and Misc. +25%	204	931	2,043
Grand Total	1,019 ft.²	4,656	10,213
Square Ft./Person	10 ft.²	9.3	10.2 ft.²
Size	22 × 46	48 × 96	71 × 143

Figure 8-10 Cafeteria Layout For 500 People

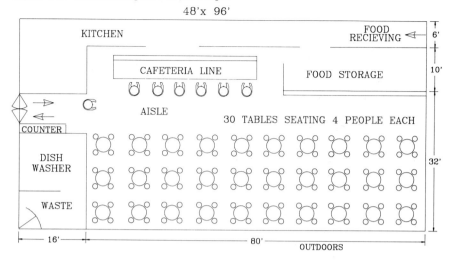

minutes will reduce waiting time or the number of facilities required in a 30-minute lunch period (15 to 20 minutes of seating is all that is used). Keep in mind that 10 to 15 minutes of employee time can be used in walking to and from the workstation and waiting in line (see Figure 8-8). A good estimate for a lunch room would be 10 square feet per employee. Figure 8-10 is an example layout for a 500 person lunch room.

RECREATION

Recreation facilities are becoming more important every year. Health conscience employees are better employees and companies are recognizing this fact. Health clubs, tracks, locker rooms, as well as ping pong tables, card games, and social clubs are becoming a part of our plants. These facilities take space, and the plant layout designer must talk with management to understand what facilities must be included. A workstation layout (like) drawing must be made for each facility, and the individual space determined and included in the plan.

DRINKING FOUNTAINS

Drinking fountains should be located within 200' of every employee and on an aisle for easy access. Each drinking fountain will include space for the drinking fountain and a person getting a drink. Fifteen square feet (3' × 5') should be allowed for each drinking fountain.

AISLES

Aisles can be the greatest consumers of plant space if we aren't careful. Aisles are for people movement, equipment movement, and material movement. The aisles must be sized for that use. For example, two-way fork truck traffic means that the aisle must be able to handle two trucks passing each other plus a safety cushion (4' + 4' + 2'). In this case, 10' aisles would be needed. Two-way people aisles must be at least 5' wide. Every workstation must have aisle access. Every shelf or rack must have aisle access.

Aisles should be long and straight. The major aisle of the plant may run from the receiving dock straight through the plant to the shipping dock. Side aisles may be smaller but perpendicular to the main aisle (see Figure 8-11).

The percentage of the plants total square footage used as aisles (square footage of aisles divided by total plant square footage) is a valuable measurement. This percentage should be plotted on a graph at least yearly. Our objective is to reduce this percentage. A couple of ideas used to reduce aisles are:

1. Use stand-up reach trucks instead of fork trucks because of their shorter turning radius.
2. Use double deep pallet racks or drive in pallet racks thereby reducing the number of aisles at least in half.

Production aisle space is allowed for in the 50 percent extra space added onto the total equipment space calculation. Warehouse aisle space is calculated with the number of aisle-feet of storage units (see Chapter 7). Access aisles around the equipment are included in the workstation layouts. Office aisles will be included in Chapter 11 with a 100 percent add-on allowance to workstation design square footage totals.

Figure 8-11 Aisle Layouts

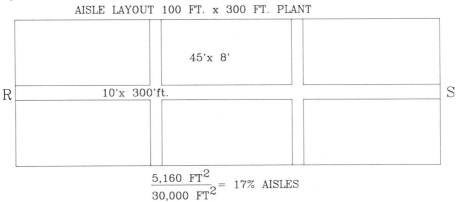

AISLE LAYOUT 100 FT. x 300 FT. PLANT

45'x 8'

R   10'x 300'ft.   S

$$\frac{5,160 \text{ FT}^2}{30,000 \text{ FT}^2} = 17\% \text{ AISLES}$$

MEDICAL FACILITIES

Medical facilities vary from 6' × 6' first aid rooms to full fledged hospitals. In smaller plants, first aid is handled by trained employees in the plant. Medical emergencies are handled by the emergency room at the local hospital or clinic. When a plant approaches 500 people, a registered nurse is usually justified. Nurses require facilities like waiting rooms, examining rooms, medical supplies, record and reclining areas. One nurse would require a 400 square foot area. If our plant has 3,000 employees, six nurses would be justified and a doctor could be hired to supervise the medical staff. Depending on the number of shifts and the type of manufacturing being done, various medical facilities will be required. Some space requirements are

Office	100 ft.²/nurse
	200 ft.²/doctor
Examining Rooms	200 ft.²/room
Waiting Area	25 ft.²/nurse and doctor
Supply Room	25 ft.²/nurse or doctor
Basic First Aid Room	36 ft.²

Figure 8-12 shows a minimum medical facility. Figure 8-13 is a layout of a larger facility.

BREAK AREAS AND LOUNGES

If the lunch room is too far away from groups of employees, a break area should be provided. "Too far" is defined as over 500' because 500' takes two minutes to walk. On a 10-minute break, two minutes walking to the lunch room and two minutes back leaves only six minutes of rest. A break area in a remote area might include a picnic table, a drinking fountain, maybe a vending machine or two, and sometimes a ping pong table that folds up and rolls away. There should be enough

Figure 8-12 Medical Facilities

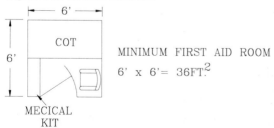

MINIMUM FIRST AID ROOM
6' x 6' = 36FT.²

Figure 8-13 Larger Medical Facility

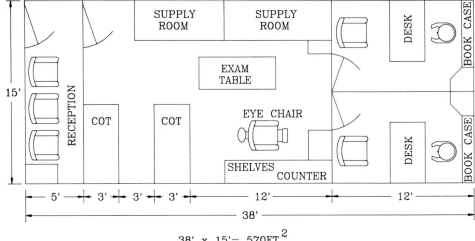

$$38' \times 15' = 570 \text{FT.}^2$$

TWO NURSE FACILITIES(4 NURSES; 2 SHIFTS)
(6 NURSES; 3 SHIFTS)

seats for everyone on breaks. Staggered breaks will reduce the need for an excessive space.

Lounges are usually found in shipping and receiving areas for visiting truck drivers to wait for their loads. The size of lounges will be proportional to the number of trucks arriving and leaving the plant each day. The lounges also keep the nonemployees (truck drivers) out of the plant. Restrooms should be conveniently close to lounges to eliminate the need for drivers walking through the plant. The lounge should be sized by multiplying the number of drivers that could be waiting at one time times 25 square feet. If we expect that four drivers would be the most drivers expected at one time, then a 100 square foot lounge would suffice.

MISCELLANEOUS EMPLOYEE SERVICES

The above list of employee services is just the main areas. You and your company will be adding to this list forever. Some of the other employee service areas are:

1. Training facilities;
2. Babysitting;
3. Barbers and hairdressers; and
4. Libraries.

See Figure 8-14 for pictures of sample employee service equipment.

Figure 8-14 Employee Service Equipment

EYE WASH

829007

829001/003

DOUBLE TIER

LINE STRIPER

Be prepared for emergencies

REST ROOM LOUNGE

797001

Courtesy of Global Equipment Co.

QUESTIONS

1. What are the employee services that require space?

2. How many parking spaces do we need?

3. How many square feet per parking space (including aisles)?
4. What is included in employee entrance space?
5. What should the employee entrance be close to?
6. What is a locker room for?
7. How big is a locker room?
8. How many restrooms do we need?
9. How do you know how many toilets, urinals, and sinks are required?
10. How big is the restroom?
11. What are the five types of eating facilities?
12. Where should the eating facility be located?
13. How big should a lunch room be?
14. How many drinking fountains should we have?
15. How big are they?
16. How much of the plants space should be taken up by aisles?
17. How many employees justify a nurse?
18. How large should a medical facility be?

Chapter 9

The Problems with Material Handling

Material handling is the function of moving the right material to the right place, at the right time, in the right amount, in sequence, and in the right position or condition to minimize production costs. The equipment to perform this function will be discussed in the next chapter. First, the principles of material handling and the control systems must be understood.

Material control systems are an integral part of modern material handling systems. Part numbering systems, location systems, inventory control systems, standardization, lot sized, order quantities, safety stocks, labeling, automatic identification techniques (bar coding) are only some of the systems required to keep industrial plants' material moving.

Material handling involves the handling equipment, the storage facilities, and the control apparatus. Material handling is also an integral part of plant layout. They cannot be separated. A change in the material handling system will change the layout, and a layout change will change the material handling system.

Material can be moved by hand or by automatic methods, material can be moved one at a time or by the thousands, material can be located in fixed location or at random, or material can be stored on the floor or high in the sky. The variations are limitless and only by cost comparison of the many alternatives will the correct answer emerge.

The proper material handling equipment choice is the answer to all our questions in this section. A material handling equipment list will include over 500 different types (classifications) of equipment, and if we multiply this number by the different models, sizes, and brand names, several thousand pieces of equipment are available for our use.

Material handling equipment has reduced the drudgery of work. It has reduced the cost of production and it has improved the quality of work life for nearly every person in industry today.

But, the handling of material is attributed to more than one half of all industrial accidents. Material handling equipment can eliminate manual lifting. Material handling equipment (like all equipment) can cause injury, so the material handling project engineers can never forget about the safety aspects.

Material handling costs, on the average 50 percent of the total operation cost. In some industries, like mining, this cost increases to 90 percent of operations cost. This fact alone justifies great effort on the part of industrial managers and technologists.

COST JUSTIFICATION

Material handling equipment can be very expensive, so all investments should be cost justified. The cheapest cost per unit gives us the best answer. If a very expensive piece of equipment reduces unit cost, it is a good purchase. If it does not reduce unit cost, it is a bad purchase.

Non-powered equipment can be very cost efficient, and should always be considered. Gravity chute, rollers, hand carts, and hand jacks are only a few of the many very popular methods of moving material economically.

Safety, quality, labor, power and equipment costs must all be included in the unit costs. If someone is expected to lift 100 pounds of parts, the time standard should allow about a 100 percent fatigue allowance built into the time standard to compensate the employee for the excessive fatigue on his/her body. When a new piece of material handling equipment eliminates this need to lift 100 pounds, the fatigue allowance should be removed from the standard to help pay for the equipment.

Sample Material Handling Cost Problem

An oil remanufacturing company uses clay in their manufacturing process. This clay comes into the plant in 80 pound bags stacked 40 per pallet and 50 pallets per boxcar. The railroad spur comes into the plant property but your plant doesn't have a rail car siding. Two car loads per year are used. The union and the company agreed that two part-time workers would be hired for one week, twice a year at the rate of $7.50 per hour to unload these boxcars. You feel this is a bad job and no one should have to work this hard. You look into this project.

Why is it done? We need the clay, and the railroad is by far the cheapest way to transport it. Let's look at it like this:

What?   = 80 pound bags of clay equals a 160,000 pound boxcar load; no other size bags are available.
Where? = From the boxcar in our yard to the storeroom which is 300′ away.
Who?   = Two temporary workers.

When? = One week, twice a year.
How? = Present method. Manually unload the pallets off the boxcar then move these pallets into the storeroom with the fork truck we already own.

This is backbreaking work, but how much could you spend improving this job? We spend one week, twice a year with two temporary employees being paid $7.50 per hour.

$$4 \text{ weeks} \times 40 \text{ hrs./week} \times \$7.50/\text{hr.} = \$1,200$$

We spend only $1,200 per year on this job. The most we could spend would be $2,400 (two years) because our company would expect at least a 50 percent return on investment. And, if we spent $2,400, we would have to eliminate all labor. There's not much chance of that. What could you buy or fabricate for $2,400? Not much! The answer remains the same. By hand is often the best way because it's the cheapest way.

GOALS OF MATERIAL HANDLING

The primary goal of material handling is to reduce unit costs of production. All other goals are subordinate to this goal. But the following subgoals are a good checklist for cost reduction:

1. Maintain or improve product quality, reduce damage, and provide for protection of materials.
2. Promote safety and improve working conditions.
3. Promote productivity:
   a. Material should flow in a straight line.
   b. Material should move as short a distance as possible.
   c. Use gravity! It's free power.
   d. Move more material at one time.
   e. Mechanize material handling.
   f. Automate material handling.
   g. Maintain or improve material handling/production ratios.
   h. Increase throughput by using automatic material handling equipment.
4. Promote increased use of facilities:
   a. Promote the use of the building cube.
   b. Purchase versatile equipment.
   c. Standardize material handling equipment.
   d. Maximize production equipment utilization using material handling feeders.

e. Maintain, and replace as needed all equipment and develop a preventive maintenance program.

f. Integrate all material handling equipment into a system.

5. Reduce tare weight (dead weight).

6. Control inventory.

THE 20 PRINCIPLES OF MATERIAL HANDLING

The College Industrial Committee on Material Handling Education, sponsored by The Material Handling Institute, Inc. and the International Material Management Society, in June of 1966, adapted the 20 principles of material handling found in Figure 9-1.

The experience of generations of material handling engineers has been summarized here for the new practitioners. These principles are guidelines for the application of sound judgment. Some principles are in conflict with others, so only the situation being designed will determine what is correct. The principles will be a good checklist for improvement opportunities. Each of these principles will be discussed in the following section.

1. The Planning Principle

General Dwight D. Eisenhower stated that the plan was nothing, but that planning was everything. The plant layout and material handling project is a plan, a drawing of where every piece of equipment goes. But General Eisenhower said the plan was nothing! What General Eisenhower was telling us is that the planning process (all that time and effort that goes into the plan) is what's important. The plan is only our way of communicating the tremendous work (planning) that went into it. Material handling planning considers every move, every storage need, and any delay in order to minimize production costs.

2. The Systems Principle

All material handling equipment should work together so that everything fits. This is the *systems concept*. The boxes fit the pallets, the pallets fit the rack, and the

Figure 9-1 The 20 Principles Of Material Handling*

1. *Planning Principle*. Plan all material handling and storage activities to obtain maximum overall operating efficiency.
2. *System Principle*. Integrate as many handling activities as is practical into a coordinated system of operations, covering vendor, receiving, storage, production, inspection, packaging, warehousing, shipping, transportation, and customer.

Figure 9-1 (cont'd) The 20 Principles Of Material Handling*

3. *Material Flow Principle.* Provide an operation sequence and equipment layout optimizing material flow.

4. *Simplification Principle.* Simplify handling by reducing, eliminating, or combining unnecessary movements and/or equipment.

5. *Gravity Principle.* Utilize gravity to move material wherever practical.

6. *Space Utilization Principle.* Make optimum utilization of building cube.

7. *Unit Size Principle.* Increase the quantity, size or weight of unit loads or flow rate.

8. *Mechanization Principle.* Mechanize handling operations.

9. *Automation Principle.* Provide automation to include production, handling, and storage functions.

10. *Equipment Selection Principle.* In selecting handling equipment consider all aspects of the material being handled—the movement and the method to be used.

11. *Standardization Principle.* Standardize handling methods as well as types and sizes of handling equipment.

12. *Adaptability Principle.* Use methods and equipment that can best perform a variety of tasks and applications where special purpose equipment is not justified.

13. *Dead Weight Principle.* Reduce ratio of dead weight of mobile handling equipment to load carried.

14. *Utilization Principle.* Plan for optimum utilization of handling equipment and manpower.

15. *Maintenance Principle.* Plan for preventative maintenance and scheduled repairs of all handling equipment.

16. *Obsolescence Principle.* Replace obsolete handling methods and equipment when more efficient methods or equipment will improve operations.

17. *Control Principle.* Use material handling activities to improve control of production inventory, and order handling.

18. *Capacity Principle.* Use handling equipment to help achieve desired production capacity.

19. *Performance Principle.* Determine effectiveness of handling performance in terms of expense per unit handled.

20. *Safety Principle.* Provide suitable methods and equipment for safe handling.

* Reprinted with the permission of the Material Handling Institute.

pallets fit the workstation. A toy company purchased parts manufactured on the outside, but these outside suppliers sent the parts into the toy company in the toy company's cartons. This company used only four different size boxes, and these boxes fit the pallets perfectly. When the parts were moved to the assembly line, the box fit into a holding device which held the box in perfect position for use.

Another example of the systems approach is a TV manufacturer. The TV manufacturer did not make the wooden cabinet but purchased it from a supplier. The supplier built the wooden cabinet, packaged it into a cardboard box which the TV manufacturer provided. The cabinet came into the TV plant, was removed from the carton, and was placed on a conveyor for the assembly of the TV set into the cabinet. The carton was then placed on an overhead belt conveyor that carried it to the packout department. When the TV was completed, it was placed back in the same carton it was received in. That carton was then moved to the warehouse and then shipped to the customers all in the same carton.

In another example, a major oil company purchased plastic bottles from an outside manufacturer. The quart bottles were packaged in a carton of 12 with separators between every bottle. These cartons were placed on a pallet and shipped into the oil company's bottling plant. In the plant, the bottles were dumped onto a filling line and filled with oil. The empty carton was conveyed to the packout end of the filling line and repacked with 12 bottles, closed, stacked on a pallet and shipped to a customer.

The systems principle integrates as many steps in the process as possible into a single system from the vendor through our plant and out to our customers. An integrated system is where everything seems to fit together.

3. The Material Flow Principle

Chapters 3, 4, and 5 discussed techniques for creating an optimum material flow layout. The fabrication analysis techniques and the assembly and packout techniques showed us how to place equipment for the shortest flow. These 10 techniques help us choose an optimum material handling system.

4. The Work Simplification Principle

Material handling, like every other area of work, should be scrutinized for cost reduction. The work simplification formula tells us to ask four questions:

1. Can this job be eliminated? This is the first question asked because a positive answer will save the maximum amount of cost, namely, everything. Material handling tasks can often be eliminated by combining production operations together.
2. *If we can't eliminate,* can we combine this movement with other movements to reduce that cost? The *unit load concept* (a special section of this chapter) is based on this work simplification principle. If we can move two for the same cost as one, the unit cost of the move will be half. Just think if we could move

1,000 instead of one. Many times moves can effectively be eliminated when combined with an automatic material handling system which moves material automatically between workstations. Conveyors are a good example of this.

3. *If we can't eliminate or combine,* can we rearrange the operations around to reduce the travel distances? Rearranging the equipment to make the travel distances less will reduce the material handling costs.

4. *If we can't eliminate, combine, or reroute,* can we simplify? Simplification is making the job easier. Transportation or material handling equipment has taken the drudgery out of work more than any other type of equipment. Some simplification ideas for material handling are:

   a. Carts instead of carrying

   b. Roller conveyors to move boxes from trucks to plant floor

   c. Two-wheel hand trucks

   d. Manipulators can make super people out of everyone

   e. Slides or chutes

   f. Roll top tables (ball bearings)

   g. Mechanization

   h. Automation

Cost reduction is a part of every engineer and manager's job. Material handling equipment makes cost reduction easier.

5. Use Gravity

Gravity power is free and there are unlimited ways to use it at workstations in order to bring material into the station and to remove finished parts. Gravity can move material between workstations. A golf club manufacturer moved golf club heads between machines on inclined skate wheel conveyors in boxes of 100. The boxes moved themselves into position at the next station. A bar stool manufacturing plant moved the finished stools away from the packout workstation with a conveyor that elevated the stool to 12′ where it dropped off onto a skate wheel conveyor and rolled up 200′ to a truck awaiting shipment.

6. Maximize the Building Cube*

A goal of material handling is to maximize the *building cube*. Racks, mezzanines, and overhead conveyors are a few of the material handling devices that promote this goal. Plant space can cost about $25 per square foot to build or $5 per square foot per year to rent. The better we use our building cube, the less space we need to buy or rent.

---

* The *building cube* is the cubic feet of the building volume resulting from multiplying the building's length times width times height.

7. Unit Size Principle

A *unit load* is a load of many parts that move as one. The advantages of a unit load is that it is faster and cheaper than moving parts one at a time. The disadvantages are:

1. Cost of making the unit loads and deunitizing.
2. Tare weight (the weight of boxes, pallets, and the like).
3. What to do with the empties.
4. The need for heavy equipment and its space needs.

Of course, the advantages must outweigh the disadvantages before we would recommend a unit load system.

The most common unit load is the pallet. Almost anything can be stacked on a pallet tied with bonding or plastic wrap and moved around the plant or world as one unit. Pallets are made of a variety of materials with greatly differing costs.

Cardboard pallets @ $1.00 each will make 1 trip.

Plastic pallets @ $4.00 each will make 20 trips.

Wooden pallets @ $20.00 each will make 100 trips.

Steel Skids @ $150.00 each will make 2,000 trips.

If we had no chance of getting our pallet or the cost of this pallet back, we would use a cardboard pallet. If we used pallets within our plant only, we would choose the steel pallet because its cost per move would be only one third of a wooden or plastic pallet. Strength, durability, versatility, weight, size, cost and ease of use must all be considered when choosing a unit load technique. Wooden pallets are the most popular because the trucking industry trades pallets. When a trucker drops of 18 full pallet loads of material, they pick up 18 empties and return them to the supplier. Tens of thousands of dollars per year can be lost without a pallet control system.

The pallet is only one of the under mass techniques of unit loading. Others exclude boxes, tubs, and slip sheets. Still others are *squeezing* and *suspending*.

Squeezing the load is performed by a clamp truck. The product is stacked on the floor into *pallet patterns* just like on pallets (see Figure 9-2). When the stack is complete, a fork truck with two vertical plates (about 4' × 4') drives up to the stack with one plate on the right side and the other plate on the left side. The two plates are pulled toward each other squeezing the material between the plates. The load can now be moved. This load can be placed on top of another stack of similar products right up to the rafters. The advantage is no pallet cost or space. Trailers can be loaded and unloaded with no pallets needed.

Suspending unit loads from bridge cranes or jib cranes. A hook suspends from a lifting motor and attaches to chains or cables around the load. Lumber, steel coils, and steel plates are often moved this way. A monorail conveyor can also move many parts at a time.

Figure 9-2 Pallet Pattern

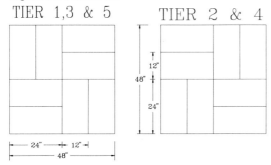

TIER 1,3 & 5   TIER 2 & 4

8. Mechanization Principle

The mechanization principle is to add power to eliminate manual moving. Mechanization is on our way to automation.

9. Automation Principle

The automation principle makes moves automatic. Many new systems are completely automatic. Automatic storage and retrieval systems place material into storage racks automatically (no people assistance) and remove it when needed. Many machines are automatic because material handling equipment load and unload the machine. Automation is the way of the future and even the manual system must think toward the future when automation will be justified.

An engine block is automatically moved from machine to machine for processing. Machine centers are arranged around an indexing table. When all machines finish their function, the table advances one station and the machines go back to work. The finished parts can be removed by gravity, or a robot can pick up the finished part and place it in a container. This principle is fun to work with because your creative efforts will be well rewarded and personally gratifying.

10. Equipment Selection Principle

Which piece of material handling equipment should we use? Which problems should be studied first? Should I do an overview before studying the individual material handling problems? These are typical questions asked by a new project engineer. Where to start is easy—just start collecting information about the product (material) and the move (job). A series of questions has been developed that has been used for generations by reporters and these questions will serve the material handling project engineer well: Why? Who? What? Where? When? How? If we answer these questions about each move, the solution will become evident.

The material handling equation is a plan for a systematic approach to equipment solution (see Figure 9-3).

Figure 9-3 The Material Handling Equation

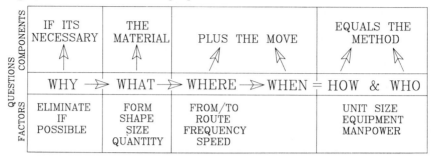

Understand the material plus the move and the proper piece of equipment will develop.

Here's a list of the specific questions to ask!

1. *Why are we making this move?* (Why?) This question is asked first because if a good answer is not forthcoming, we can eliminate this move. By combining operations, the move between operations can be eliminated. We can combine machines together (called *work cells*) and eliminate moves.

2. *What are we moving?* (What?) Understanding what is being moved requires a knowledge of size, shape, weight, number being moved and the kind of material. With the knowledge of what needs to be moved, we have half the information required to make an equipment selection.

3. *Where are we moving the material from and to?* (Where?) If the move is the same every time, a fixed path technique is warranted (conveyor). If the move changes from part to part, a valuable path technique is used (industrial truck). If the path is short, maybe gravity can be used (slides, rollers, skatewheel).

4. *When is the move needed?* (When?) Is this move once or twice a day? If so, an industrial truck is warranted. If this is several times a minute, a conveyor is used. A few examples of analyzing the material + move = method follow #5 below.

5. *How will the move be made?* (How?) Will we move it by hand, or by conveyor, or by fork truck. Many options are available to us and the cheapest unit cost is our goal.

Here are a few examples of ways of putting these questions in action.

Example 1: Oil from a tanker truck to a tank farm for a bottling company.
  Why = we need oil to put into quart cans
  What = oil
Where = from tank truck to tank farm
  Who = receiving clerk

When = 4 times a day as they arrive
How = pump—meter and hose

Example 2: Unload 20,000 pounds of flat steel stock from a flatbed trailer into our plant.

Why = we need steel in our plant (maybe coil would be better)
What = 20,000 pound loads of steel (3-1/2' × 10' × 20")
Where = from flatbed trailer into our storage area
Who = receiving clerk
When = 40,000 pounds per day (one truck)
How = bridge crane

Example 3: Move parts one at a time from spotweld through paint to assembly.

Why = Moving a major amount of our product automatically
What = Tool boxes and tote trays
Where = From spotweld to paint to assembly
Who = Automatically
When = 11 parts per minute, 432 minutes per shift
How = Overhead conveyor belt

Example 4: Move all product (except sheet metal) from the receiving dock to stores and/or to manufacturing.

Why = to keep a bank of material so we don't run out of raw material and parts
What = all raw material and purchased parts
Where = from receiving to stores to manufacturing
Who = store clerk
When = as material arrives and as production requests material
How = narrow aisle reach truck

As information is collected the picture becomes clearer and the plan takes shape. The more you know about the material and the move, the better the job of equipment selection you will do.

11. The Standardization Principle

There are many types of material handling equipment—shop boxes, bins, pallets, shelves, racks, conveyors, trucks, and the like—and in every area we want to standardize on one (or as few as possible) size, type, and even brand name. The reasons are many, and the reasons change with the type of equipment, but if we have a special piece of equipment for every move or storage, we will have too many different types and sizes to inventory and control. Material handling moving equipment (like fork trucks) are manufactured by many companies. We need to choose just one and then stay with that brand, type and size forever because spare parts inventory, maintenance, and operation of this equipment will be most cost efficient. If a new technologist were to recommend a new brand fork truck because its on sale, management should reject the recommendation because of the expense of training operators, training maintenance people and maintaining an inventory of new spare parts.

Having only a few sizes of cartons will simplify the storage area. We may put these few sizes of cartons onto a single sized pallet and into a uniform sized rack which is serviced by one type of lift truck.

12. The Adaptability Principle

Use equipment that can do many different jobs without excessive changeover time or costs. If special purpose equipment can be justified in a reasonable amount of time, then go ahead, but remember that change is inevitable, and your special purpose equipment could be obsolete and useless. The adaptability principle is the best reason to buy a fork truck which is very versatile. With any amount of production volume, there is almost always a better way to move material other than by fork truck.

Buy standard size pallets, buy shop containers that will handle a variety of parts, buy storage equipment that can store a wide variety of products. In this way, change will be less costly.

13. The Dead Weight Principle

"Don't use a 20 pound sledge hammer to drive a tack." Try to reduce the ratio of equipment weight to product weight. Don't buy equipment that is bigger than necessary.

Tare weight is a term used to describe the weight of packaging material. When we move product, we place the product into a container, we may place packaging materials around the product to prevent damage while moving. These containers may be placed on pallets as well. The container, the stuffing and the pallet all add up to the tare weight. If we ship this packaging, the tare weight costs as much to ship as does our product. This packaging is also expensive to buy. So, the goal is to reduce this tare weight and save money.

14. The Utilization Principle

Material handling equipment and operators should be used fully. Knowing what work is required, the number of times per day, and the time required per move will help us manage the workload of our people and the equipment.

15. The Maintenance Principle

Material handling equipment must be maintained. Preventative maintenance (periodic, planned maintenance) is cheaper than emergency maintenance, so a preventative maintenance program including schedules must be developed for each piece of material handling equipment.

Pallets, shop boxes, and storage facilities need repair too. Missing slats on pallets can cause product damage and safety problems. Wooden pallets cost about $20 each, so we don't want to throw them away because one slat is broken. Set up pallet repair area to store and repair broken pallets.

16. The Obsolescence Principle

As equipment wears out or a better, more efficient method becomes available, let's replace that equipment and improve our operation. Good maintenance records will help us identify worn out equipment. Good technologists are always looking for better ways to improve our operations.

17. The Control Principle

Materials are costly and material handling systems can be a part of the inventory control system. Conveyors can move material past a scanner to count, identify, and reroute.

18. The Capacity Principle

We want to get as much out of our production equipment and our employees as possible. Material handling equipment can help maximize production equipment utilization.

A punch press can cycle every .030 minutes of 33 times per minute, but our time standard for manually loading and unloading this press is only 300 pieces per hour. This is only 15 percent of the machine's capability.

$$\frac{60 \text{ min. per hr.}}{.030 \text{ min per unit}} = 2,000 \text{ pieces/hr. potential}$$

$$\frac{300 \text{ pieces per hr. present}}{2,000 \text{ pieces/hr. potential}} = 15\%$$

If we purchased a coil feeding material handling system, we would approach 100 percent machine utilization.

Material handling equipment can assist our production equipment in achieving its potential. Don't buy a new machine, just get the capacity available in our present machine.

19. The Performance Principle

Know what your material handling costs are and work at reducing them. The process chart, discussed earlier in this book, gives us a form to calculate the unit cost of every move. This is a starting point for cost reduction. Material handling labor moves material, and a measurement of output could be pounds moved. Input is labor hours. Anything we can do to increase pounds moved or reduce labor hours will increase productivity.

Performance of material handling can also be calculated by ratios.

$$\% \text{ of material handling} = \frac{\text{material handling hrs.}}{\text{total labor hrs.}}$$

Tracking this percentage can show the improvements in performance.

Performance includes a lot more than labor. Segregating material handling cost from total operation costs would result in a better ratio. Again, improvement in the ratio would indicate improved performance.

20. The Safety Principle

Manual handling is probably the most dangerous method of material handling, and as stated earlier, material handling equipment has improved the world of work more than any other area of industry. Material handling equipment can also be a source of safety problems, so safety methods, procedures and training must be a part of any material handling plan. It is managements responsibility to provide a safe work environment. Tens of billions of dollars are spent on injured workers. This human cost is also reflected in the cost of our product. We must eliminate safety problems, and material handling equipment can play a major role in this goal. The U.S. Bureau of Labor Statistics (BLS) data recorded in 1988 showed 90 million workers in the United States and 6.2 million job-related injuries.

The National Council on Compensation Insurance says we spent $40 billion on medical treatment, lost wages, and death benefits in 1988. Eleven billion dollars was spent on lower back injuries alone. Lower back injuries account for 37 percent of all disabling occupational injuries and nearly 20 percent of all workers' compensation claims are filed for back injuries.

THE MATERIAL HANDLING PROBLEM SOLVING PROCEDURE

Step 1: Analyze the requirements to define the problem. Be sure the move is required.

Step 2: Determine the magnitude of the problem. Cost analysis is best.

Step 3: Collect as much information as possible—why, who, what, where, when, and how.

Step 4: Search for vendors. Suppliers often provide outstanding engineering and cost justification assistance.

Step 5: Develop viable alternatives.

Step 6: Collect costs and savings data on all alternatives.

Step 7: Select the best method.

Step 8: Select a supplier.

Step 9: Prepare the cost justification.

Step 10: Prepare formal report.

Step 11: Make presentation to management.

Step 12: Obtain approvals (adjust as needed).

Step 13: Place order.

Step 14: Receive and install equipment.

Step 15: Train employees.

Step 16: Debug (make it work) and revise as necessary.

Step 17: Place into production.

Step 18: Follow-up to see that it's working as planned.

Step 19: Audit performance to see that payback was realized.

MATERIAL HANDLING CHECKLIST

100 Areas of Cost Reduction

Yes	No	
___	___	1. Is the receiving and shipping docks protected from the weather?
___	___	2. Are dock boards adequate?
___	___	3. Do you load and unload trailer by hand?
___	___	4. Is your incoming material packaged for your economical use?
___	___	5. Do you replace obsolete equipment?
___	___	6. Do you standardize on material handling equipment to reduce spare parts needs?
___	___	7. Do you have a preventative maintenance program for every piece of material handling equipment?
___	___	8. Do you have a pallet repair area?
___	___	9. Do you have a pallet control program?
___	___	10. Do you measure and track the ratio of material handlers to direct labor?
___	___	11. Do you have a material handling training program?
___	___	12. Do you maintain safety records for material handling?
___	___	13. Do any of your employees lift 50 pounds or more manually?
___	___	14. Are there any material handling jobs that require more than one person to lift?
		15. Are you using the space overhead in:
___	___	a) stores?
___	___	b) fabrication?
___	___	c) paint?
___	___	d) assembly and packout?
___	___	e) warehouse?
___	___	f) offices?
___	___	16. Is weight control measured and recorded automatically?
___	___	17. Do you still receive raw material (like plastics) in 50 to 100 pound bags when your usage would justify bulk handling equipment?
___	___	18. Are you storing material that is available locally?
___	___	19. Are you using the building cube?
___	___	a) Are you storing only 8' high?
___	___	b) Are you picking orders only 6" high?
___	___	c) Are you using overhead conveyors?
___	___	d) Are your ovens off the floor?
___	___	e) Are you using over aisle storage?
___	___	f) Are you stacking two or more deep?
___	___	g) Is more than 30 percent of your plant taken up by aisles?
___	___	20. Are you using powered equipment when gravity would do the job?

Yes	No	
——	——	21. Do you use the material handling equipment to do secondary operations automatically?
——	——	a) counting
——	——	b) weighing
——	——	c) brand or number
——	——	d) segregate
——	——	e) slit bags
——	——	f) open and close lids
——	——	g) glue boxes closed
——	——	h) band boxes
——	——	22. Do you automatically move material to point of use, then hand feed?
——	——	23. Is the maintenance and service area for mobile equipment conveniently located?
——	——	24. Are skilled employees spending their time handling material?
——	——	25. Does the assembly line stop when delivering and removing material?
——	——	26. Do operators have to load their own hoppers?
——	——	27. Do operators need to stop work when material is being loaded into their workstation?
——	——	28. Are material storage areas congested?
——	——	29. Do you measure the utilization of material handling equipment?
——	——	30. Do you encourage backhauling?
——	——	31. Does your equipment move empty more than 20 percent of the time?
——	——	32. Do your shipping clerks load carrier's trucks?
		33. How do you load material handlers with work?
——	——	a) by past practices
——	——	b) by time standards
——	——	c) guess or no thought
——	——	34. Do you pay demurrage charges?
——	——	35. Do you know your floor loadings?
——	——	36. Does your product get damaged during material handling?
——	——	a) Do you know which equipment?
——	——	b) Do you know how much ($)?
——	——	c) Do you know which people?
——	——	37. Do you use two way radio for truck drivers?
——	——	38. Do you handle material too many times?
——	——	39. Are single pieces being moved where two or more could be moved?
——	——	40. Are your floors smooth and clean?
——	——	41. Do you know the capacity (pounds) of your equipment?
——	——	a) Do your material handlers know?
——	——	b) Is your equipment marked for capacity?
——	——	42. Do you ever change material from one container to another?
——	——	43. Are your aisles over 8'?
——	——	44. Have you flow process charted your product?
——	——	45. Do you use certified truck load weight?
——	——	46. Do you use point of use receiving?
——	——	47. Do you ever store material in aisles?
——	——	48. Do you use outdoor storage when practical?
——	——	49. Is your safety stock too large?
——	——	50. Are your doors too small?
——	——	51. Do you have too many doors?
——	——	52. Do you control the movement of material?
——	——	53. Do you have a locator system?

Yes	No	
_____	_____	54. Do you use drums instead of tanks?
_____	_____	55. Do you use trailable trailers for long hauls in the plant?
_____	_____	56. Are you ground loading or unloading trailers or boxcars?
_____	_____	57. Can your customers unload your trailers?
_____	_____	58. Can we eliminate pallets?
_____	_____	59. Can we use expendable pallets (one way)?
_____	_____	60. Can we build material handling devices into our finished package?
_____	_____	61. Are we using fork trucks where narrow aisle trucks would be better?
_____	_____	62. Do you use the receiving container to ship?
_____	_____	63. Do you have material handlers waiting on assembly and packout liner?
_____	_____	64. Have you eliminated backtracking?
_____	_____	65. Have you eliminated cross traffic?
_____	_____	66. Have you reduced the distance of travel to a minimum?
_____	_____	67. Are materials too far from their point of use?
_____	_____	68. Are the containers at a workstation large enough to keep the station working for two hours or more?
_____	_____	69. Are inexpensive parts stored clsoe to the workstation?
_____	_____	70. Is scrap disposal considered?
_____	_____	a. separation
_____	_____	b. storage area
_____	_____	c. sold (not given away)
_____	_____	71. Can the shipping container start the production line?
_____	_____	72. Are we bulk storing finished goods to maximize building cube?
_____	_____	73. Can turntables eliminate steps?
_____	_____	74. Do you use light signals to notify material handlers of stock needs?
_____	_____	75. Do we have the high use items conveniently located?
_____	_____	76. Have you done an A B C analysis of inventory?
_____	_____	77. Are we breaking carton quantities for shipments?
_____	_____	78. Are we using hydraulic hand carts for short trips?
_____	_____	79. Are our maintenance people mobile?
_____	_____	80. Are you continually looking for ways to mechanize?
_____	_____	81. Do you ask operator's opinion on what would make their jobs easier?
_____	_____	82. When an employee's suggestion is rejected, do you formally tell the employee why?
_____	_____	83. Do you listen to material handling vendors (salespeople)?
_____	_____	84. Have you asked vendors to package raw materials more conveniently?
_____	_____	85. Do you prequalify vendors so that their material will not need inspection upon receipt?
_____	_____	86. Do you have the best systems for mailing, strapping, taping, stapling, labeling, and marking your product?
_____	_____	87. Are we using conveyors as much as possible?
_____	_____	a) receiving to stores?
_____	_____	b) stores to production?
_____	_____	c) fabrication through clean, paint and bake to assembly?
_____	_____	d) between operations?
_____	_____	e) through heat treatment?
_____	_____	88. Are you loading machines with walking beams?
_____	_____	89. Do you use indexing tables?
_____	_____	90. Are you using computer generated shipping labels?
_____	_____	91. Are you generating computer weights for shipments?
_____	_____	92. Are you using computer location/numbered layouts for warehousing?
_____	_____	93. Are all doors equipped with automatic approach openers?

MATERIAL HANDLING CHECKLIST

Yes	No	
_____	_____	94. Are corners protected with signals and mirrors?
_____	_____	95. Are you using remote controllers and indicators to eliminate climbing and walking to remote areas?
_____	_____	96. Is equipment protected with panic buttons?
_____	_____	97. Are switches foot, knee or leg controlled so hands are free to work?
_____	_____	98. Are you using air, electric, or hydraulic clamps to eliminate the need for hand holding?
_____	_____	99. Are jigs being used to hold parts in position for welding, cementing or assembly?
_____	_____	100. Are we using hand held power tools instead of hard tools?

QUESTIONS

1. What is material handling?

2. What are some of the material control systems?

3. Which is the best piece of material handling equipment for a specific job?

4. What are the 18 goals of material handling?

5. Where did the 20 principles of material handling come from?

6. What are the 20 principles of material handling?

7. What is the material handling equation?

8. What is the material handling problem solving procedure?

Chapter 10

Material Handling Equipment

How do we choose the proper piece of equipment from the thousands of material handling devices available to us? For the experienced project engineer or manager, this problem is not as great as for the new technologist. To assist the new technologist, the following organization of material handling equipment is suggested. This organization follows the flow of material from the receipt of material to the warehousing of that material as follows:

1. Receiving and shipping (because they are similar)
2. Stores
3. Fabrication
4. Assembly
5. Packaging
6. Warehousing

This organization lends itself to specific problem-solving situations. Two additional areas of material handling need to be discussed because of their importance:

7. Bulk material handling
8. Automatic storage and retrieval systems

The systems principle of material handling states that material handling devices should be used in as many areas as possible, and that everything fits (works) together. In the following discussion of material handling equipment, each piece of equipment will be discussed in the first area of use or the most common area of use.

Receiving and shipping material handling equipment are often the same. Sometimes, receiving and shipping are accomplished through the same dock door. To save time, both these important departments will be discussed at the same time.

Receiving and Shipping Docks

Receiving and shipping docks come in a variety of sizes and shapes. The term ''dock'' comes from the shipping industry where ships pulled into port, landed, were tied up and unloaded. Our industrial plants' docks are for the same purpose. Trucks, trains, and ships all can pull into our plants to deliver or remove materials.

The most common type of dock is known as a *flush dock*. Flush docks are doors in an outside wall. The truck or train pull up or back into the doorway. Trucks are serviced from the rear door mostly, but some trucks and most boxcars are serviced from the sides. Flush docks can be designed for both rear and side service. The height of the plant floor off the driveway or railbed should be 46″ for trucks and 54″ for boxcars. The driveways should slope away from the plant to prevent water damage to building foundations. Docks that slope toward the plant are common after-thought docks or short-term cheap construction. The negative slope docks will be continuing problems even with drains. Everything seems to blow into and collect in the wells. Figure 10-1a shows a side view of a flush rear service dock. Notice the dock height, slope away from plant, bumpers, chalks, and dock plate. Notice also the front wheels of the trailer. These small wheels will place a lot of weight on the driveway. Be sure the driveway can support the heavy weights of trailers. Figure 10-1b is a top view of the same dock. Figure 10-1c is a top view of a side service dock. This could have been just as easily a railroad and boxcar.

Figure 10-1a Flush Dock—Side View

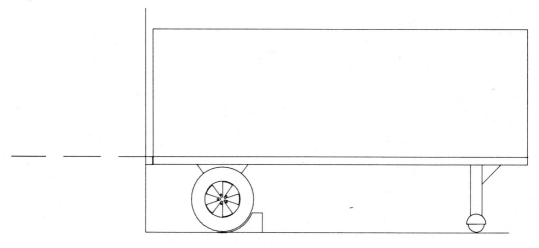

CHAP. 10: MATERIAL HANDLING EQUIPMENT

Figure 10-1b Flush Dock—Top View

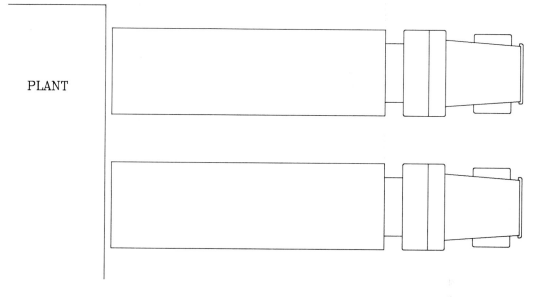

PLANT

Figure 10-1c Flush Dock—Side Service Top View

Drive-in docks are just like flush docks except that when the door opens, the truck and trailer can back into the plant and the door can be closed. This type of dock is very expensive and very desirable in bad weather. Drive-in docks can service rear end and side both. Some large plants have railroad siding where trains back cars into the plant. Caterpillar Tractor Company actually drive their new products from the plant floor straight across to the deck of a flat car.

Drive-through docks are a pair of doors across the plant from each other. The truck drives into the plant. The door is closed, the trailer (usually flat bed) is unloaded or loaded, the other door is opened, and the truck drives out. The driveway and the plant floor are usually at the same level, so personnel must climb on the trailer to unload. Overhead bridge cranes are often used to unload very heavy pallets of steel. This type of dock is a bigger safety hazard than the other docks, but good step ladders and operator training will minimize the danger.

Finger docks are extensions to the plant to handle many trailers at one time. A concrete finger from 10 to 15′ wide can be extended one hundred feet out from a side of the plant. Ten trailers can now be backed up to both sides of this dock, and only one or two doors go into the plant. Freight companies use this type of dock to unload and load many trailers at the same time. Figure 10-2a is a drawing of a finger dock for 20 trucks. A carport like cover over the finger dock is important for weather protection.

There are other docks, some good, some bad, but these four types of docks represent over 90 percent of all docks for trucks or rail cars.

Figure 10-2α Finger Dock—Top View

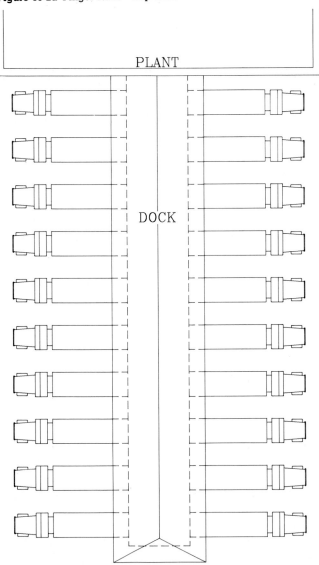

Figure 10-2b Side View

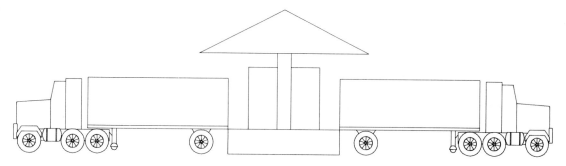

Dock Equipment

A door is a part of a dock (see Figure 10-3a). A trailer (end service) dock door is typically 9′ × 9′ rollup. The door is electrically or manually lifted into a container mounted above the door opening. Around the door, a seal is built of compressible material which will seal off the outside weather from the plant environment. Bumpers are placed outside the dock door below the floor level to stop the trailer and protect both the building and the trailer from collision damage. Once the trailer is in position and the doors of the plant and the trailer are open, a dock plate is placed between the plant floor and the trailer floor to enable us to drive on and off the trailer. An automatic dock leveler can be built into the dock door floor to make placing the dock plate automatic.

An awning or porch over the rear of the trailer extending from the plant wall will help keep rain and snow out of the plant. Sometimes air curtains and plastic curtains are placed in doorways to minimize air loss from the plant.

Extra lighting is often needed inside trailers; so portable lighting is often useful. Figure 10-3b is a collection of dock equipment pictures.

Moving Equipment

Hand carts. Literally hundreds of different hand carts are available today. A few of the most versatile and popular pieces of equipment follow.

1. Two-wheel hand jack (see Figure 10-4). Weights up to 500 pounds can be moved by a single person and good two-wheel hand jack. Hand jacks are used in about every area of business, even in the office.

2. Pallet hand jack or pallet truck-hydraulic lift (also called just hand jacks). (See Figure 10-5.) Hand jacks are rolled under a pallet, the handle is pumped (hydraulic pump handle), the pallet is lifted off the floor a few inches, and now the pallet and up to 2,000 pounds of material can be moved easily by hand.

3. Four-wheel hand trucks (see Figure 10-6). There are hundreds of shapes, sizes and uses of hand carts. You can build any shape on the platforms and move very special material. The examples in Figure 10-6 are very versatile. Many things can be loaded and moved nearly anywhere.

4. Pallets (see Figures 10-7 and 10-8). Pallets have been discussed in Chapter 9, but it is an important piece of material handling equipment.

Figure 10-3a Dock Equipment

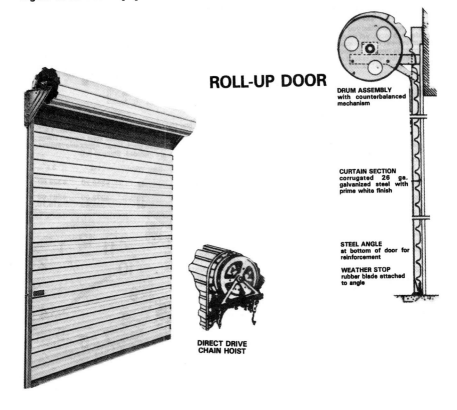

ROLL-UP DOOR

DRUM ASSEMBLY
with counterbalanced
mechanism

CURTAIN SECTION
corrugated 26 ga.
galvanized steel with
prime white finish

STEEL ANGLE
at bottom of door for
reinforcement

WEATHER STOP
rubber blade attached
to angle

DIRECT DRIVE
CHAIN HOIST

Courtesy of Global Equipment Co.

Figure 10-3b Dock Equipment

DOCK BOARD

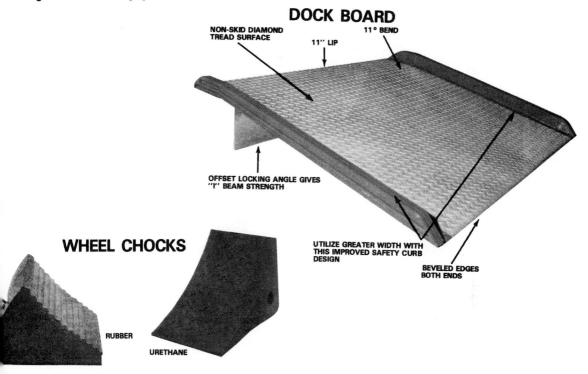

NON-SKID DIAMOND TREAD SURFACE

11" LIP

11° BEND

OFFSET LOCKING ANGLE GIVES "I" BEAM STRENGTH

UTILIZE GREATER WIDTH WITH THIS IMPROVED SAFETY CURB DESIGN

BEVELED EDGES BOTH ENDS

WHEEL CHOCKS

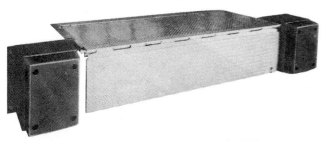

RUBBER

URETHANE

EDGE-OF-DOCK LEVELER
20,000 lb. capacity

DOCK BUMPER

Courtesy of Global Equipment Co.

Figure 10-4 Two-wheel Hand Trucks

Figure 10-5 Hand Pallet Trucks

Courtesy of Monroe Equipment Co.

Courtesy of Global Equipment Co.

Figure 10-6 Four-Wheel Hand Carts

Courtesy of Global Equipment Co.

Figure 10-7 Pallets in Pallet Rack

Courtesy of Global Equipment Co.

Fork Trucks

Fork trucks are by far the most popular piece of material handling equipment for unloading and loading trucks and rail cars (see Figure 10-9). Fork trucks are versatile. They can go about anywhere and they can move anything. Every department in the plant can use them, but they are probably the most misused of all equipment in the plant. It's too much equipment moving too little weight. There is almost always a better choice than a fork truck. Fork trucks have only one redeeming quality—they are versatile. In the stores department, we will talk

Figure 10-8 Pallets

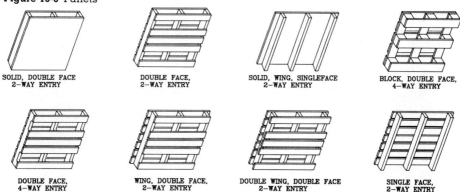

| SOLID, DOUBLE FACE 2–WAY ENTRY | DOUBLE FACE, 2–WAY ENTRY | SOLID, WING, SINGLEFACE 2–WAY ENTRY | BLOCK, DOUBLE FACE, 4–WAY ENTRY |

| DOUBLE FACE, 4–WAY ENTRY | WING, DOUBLE FACE, 2–WAY ENTRY | DOUBLE WING, DOUBLE FACE 2–WAY ENTRY | SINGLE FACE, 2–WAY ENTRY |

Figure 10-9a Industrial Trucks

Courtesy of Yale Materials Handling Corp.

Figure 10-9b Fork Truck

Courtesy of Crown Equipment Corp.

Figure 10-9c Fork Truck

Courtesy of Yale Materials Handling Corp.

about narrow aisle trucks that can turn in a much shorter radius and the operator can get off and on easier and faster.

Fork truck attachments are available for more specific jobs. Standard forks are not appropriate for moving paper rolls, carpet rolls, drums, trash, or many other parts and containers, but fork truck attachments are available to create a very unique lifting and moving device. A boat marina uses extended forks to put motor boats into a storage rack. Oil, paint, scrap and parts can be dumped by using a special dumping attachment. Figure 10-10 show a few attachments.

Bridge Cranes

Bridge cranes (see Figure 10-11a and b) are so named because they bridge a bay (wall to wall). Columns are placed at intervals, lets say, 40' × 60'. The bay will be 60' wide. Two rails (like a big railroad track) are mounted to the columns and can run the full length of the bay and even outside at 60' apart. The bridge then runs on wheels on these two tracks. On the bridge, a lifting motor is attached. This lifting motor travels back and forth under the bridge. The crane can be operated from the floor or, on bigger units, an operator rides in a cab mounted to the bridge.

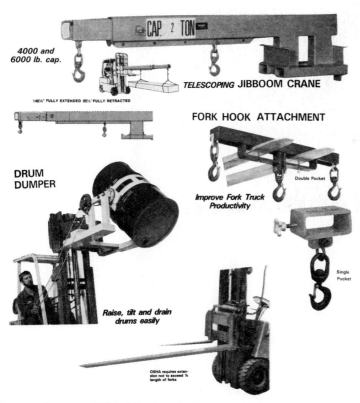

Courtesy of Global Equipment Co.

Figure 10-11a Bridge Crane

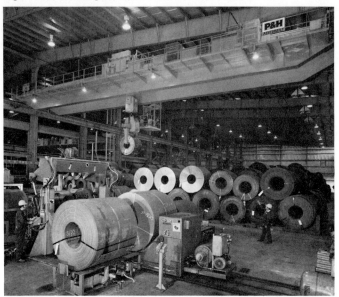

Figure 10-11b Bridge Crane

Courtesy of Harnischfeger Corporation,
Manufacturers of Overhead Lifting Equipment.

Bridge cranes can lift and move very heavy loads—up to 100,000 pounds or more. Steel, bar stock, major subassemblies, and the like are moved with bridge cranes.

Bridge cranes mounted on the wall on one side and a leg on the other side are called *single gantry cranes*. For outside use, both ends are attached to legs, this is a double gantry crane. Cargo ships are loaded using very large double gantry cranes (see Figure 10-11c).

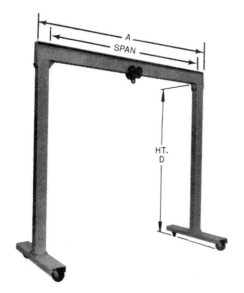

Figure 10-11c Double Gantry Crane (A-Frame)

Courtesy of Air Technical Industries.

Telescopic Conveyor

Telescopic conveyors have several sections of conveyor that extend as needed (see Figure 10-12). When unloading a truck of small cartons, the first cartons are next to the door, but as the truck is unloaded, the cartons get further and further away from the door. A telescopic conveyor can move into the trailer as the work dictates. Telescopic conveyors can save many feet of travel. The Sears warehouses receive much of their incoming goods by way of telescopic conveyor. One conveyor covers two dock doors. The telescopic conveyors are connected to additional flat belt conveyors, but flat belt conveyors will be talked about in the assembly department section of this chapter because belt conveyor almost describes assembly material handling.

Weight Scale

Weight scales are valuable receiving and shipping tools that are built into the material handling system (see Figures 10-13a and 10-13b). On receiving docks, scales are used for counting incoming material. Drive on scales, where fork trucks place a pallet of material onto the scale to weigh it before shipping. Pedestal scales can be built into a conveyor line to weigh material automatically. Weight scales assist in the quality control of receiving and shipping counts.

RECEIVING AND SHIPPING

Figure 10-12 Telescopic Conveyor

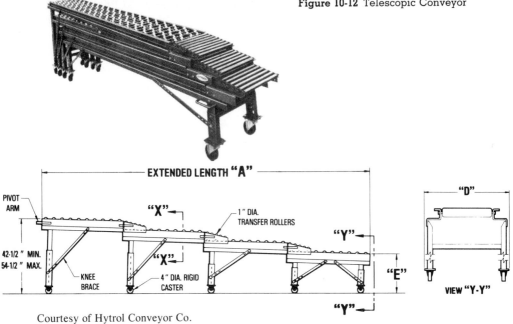

EXTENDED LENGTH **"A"**

PIVOT ARM

"X"

1" DIA. TRANSFER ROLLERS

"Y"

42-1/2 " MIN.
54-1/2 " MAX.

"X"

KNEE BRACE

4" DIA. RIGID CASTER

"E"

"Y"

"D"

VIEW "Y-Y"

Courtesy of Hytrol Conveyor Co.

Figure 10-13a Weight Scale

*VERTEX Floor Scale
High Accuracy
Option*

Many installations (batching, filling, counting, etc.) require better accuracy and repeatability than a standard commercial floor scale can provide. VERTEX High Accuracy scales meet these needs. They're designed to provide the added performance required for these applications.

■ One part in 10,000 accuracy. (Not legal for trade)
■ Includes a Factory Test Loading Report verifying 10,000 division accuracy.
■ Ideal for counting, inventory, batching, filling, etc.

VERTEX

Courtesy of Toledo Scales.

Figure 10-13b Low-Profile Scale

LOW-PROFILE SCALE
• *5000 lb. capacity*
• *Accurate to 1 part in 5000*

CRANE SCALE

DIGITAL SCALE

LEGAL FOR TRADE

Courtesy of Global Equipment Co.

Systems Required on Shipping and Receiving Docks

Systems on the shipping and receiving docks should include:

1. Part numbering systems allow for identification of inventory.
2. Purchase order system authorizes the receiving of material (see Figure 11-12).
3. Customer order system authorizes the shipment of material.
4. Bill of lading authorizes a trucking company to move our material and to bill us for their services.

STORES

"Stores" is the term used to describe the room where we hold raw materials and supplies until they are needed by the operations department. The raw material stores is usually the largest, but maintenance and office supplies stores can be large as well. The material handling equipment in stores area tend to be very expensive.

Storage Units

Storage units can include:

1. *Shelves*—Small parts are stored on shelving. A typical shelving unit looks like a book shelf with (6) $1' \times 1' \times 3'$ shelves one over the other (see Figure 10-14a).

Figure 10-14a Industrial Shelves

Courtesy of Yale Industrial Trucks.

2. *Racks*—Palletized material is generally stored on pallet racks. A typical pallet rack is 9' wide with five tiers for a height of 22'. Two pallets per tier, five tiers high equals 10 pallets per pallet rack (see Figure 10-14b).

Figure 10-14b Pallet Racks

Courtesy of Global Industrial Equipment.

3. *Double deep pallet racks* allow for stacking twenty pallets on both sides of the aisle instead of 10 pallets. The density of storage is much better, and utilization of building cube is improved (see Figure 10-14c).

4. *Portable racks* are racks that fit over a pallet load of soft material. Another pallet is then set on top of this portable rack. Heights can be much higher without the danger of a stack falling over (see Figure 10-14d).

Figure 10-14c Double Deep Pallet Racks

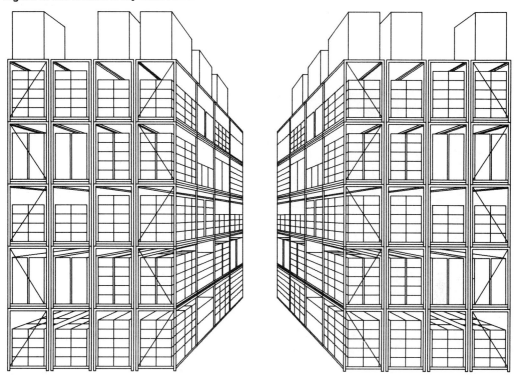

Figure 10-14c (cont'd) Double Deep Pallet Rack

Courtesy of Ridge Rak, Inc.

Figure 10-14d Portable Rack

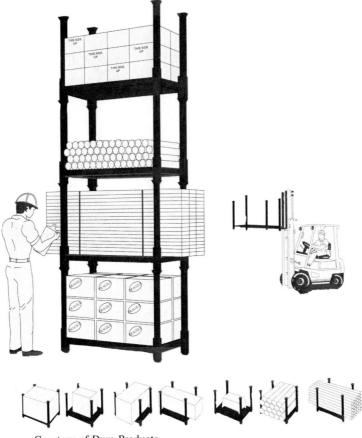

Courtesy of Dura Products.

5. *Mezzanines* can be built over shelving areas to use the space over the shelves. Additional shelves can be placed on the mezzanine doubling the number of shelves in the stores area. Slow moving stock can be placed on the mezzanines (see Figure 10-14e).

6. Rolling shelves allow for only one aisle in maybe 10 rows of shelves. This would save 9 aisles. The shelves are on wheels and tracks. The shelves can be moved to open up an aisle where no aisle exists now. Rolling shelves are popular in maintenance stores and office supply stores (see Figure 10-14f).

7. Drawer storage units are another popular maintenance storage unit because many small parts can be stored in a small area. One drawer may have 32–64 storage locations, and a six foot drawer unit could hold nearly 1,000 different parts (see Figure 10-14g).

192   CHAP. 10: MATERIAL HANDLING EQUIPMENT

Figure 10-14e Mezzanines

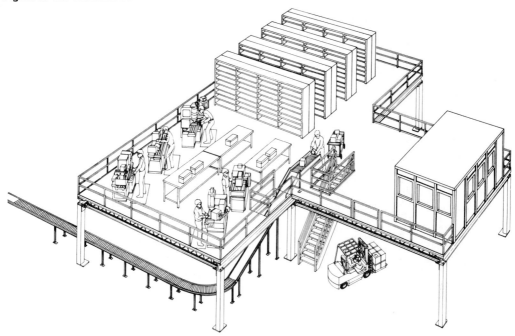

Courtesy of W.A. Schmidt, Inc.

Figure 10-14f Rolling Shelves

Courtesy of Acme Visible Records.

Figure 10-14g Drawer Storage

Courtesy of Global Equipment Co.

Stores Mobile Equipment

Narrow aisle trucks are one of those better choices when thinking of a fork truck. Narrow aisle trucks can turn in narrow aisles, and the operator stands up. Both of these features increase productivity of the companies resources:

1. Narrow aisles save space.
2. Stand up save operator time and makes getting off and on very easy. A fork truck driver sits up in the air about 3 to 4'. Once the operator gets seated up there, he/she doesn't come down except for lunch and breaks.

There are several types of narrow aisle trucks:

1. *Reach truck*—A reach truck (narrow aisle reach truck) has a scissor attachment on the forks allowing them to be extended over 4' (see Figure 10-15). This allows for the stacking of two pallets deep in an 8' deep rack. Two pallets deep will save about 50 percent of the aisle space.
2. *Straddle truck*—The name (straddle) comes from the trucks ability to straddle a pallet on the floor—one front leg on both sides of the pallet on the floor (see Figure 10-16). This allows for more stability and the ability to lift heavier loads with lighter weight vehicles.

Figure 10-15 A Reach Truck

Figure 10-16 A Straddle Truck

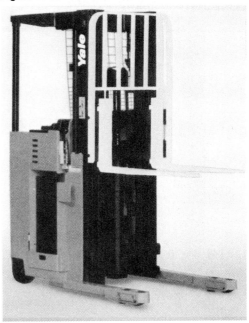

Courtesy of Yale Materials Handling Corp.

Courtesy of Yale Materials Handling Corp.

3. *Side shifting lift trucks*—Narrow aisle side shifting lift trucks are the most space conserving mobile equipment for storerooms (see Figure 10-17). Many different sizes and shapes of this class exist, but one of the most useful and unique is used for bar stock of 10 to 20′ of length. How else could be move very long lengths. This bar stock would be stored in a cantilever rack.

4. *Maintenance carts*—Maintenance carts are almost unique to the maintenance person (see Figure 10-18). Portable oil and grease carts, portable welders, portable tool boxes and portable benches are all possible. The purpose and goal of maintenance carts is to eliminate the need to run back and forth from the maintenance department to the problem because something was forgotten. The cart is a small maintenance storeroom.

Figure 10-17 Side Shifting Truck

Courtesy of Crown Equipment Co.

Figure 10-18 Maintenance Cart

Figure 10-19 Dollies and Casters

Courtesy of Global Equipment Co.

5. *Dollies and casters*—Moving equipment is a common maintenance job. Dollies placed under the equipment can expedite things (see Figure 10-19). For moving a desk, for example, one would use a dolly that looks like a pallet with wheels.

6. *Maintenance tool crib*—This is used for the safekeeping of maintenance tools and supplies (see Figure 10-20).

7. *Carousel storage and retrieval systems*—Visualize the conveyor system at your dry cleaners (see Figure 10-21). When you come in to pick up your cleaning, the clerk pushes a botton and a conveyor moves all the clothes until yours arives. In a parts storeroom, the same efficient system can be achieved by using a carousel conveyor. Each bin is numbered, and when we put something away, we make a record of that item and the bin number where we stored that part. When needed, we look up the part number we need and find its location number. Push the button and the part comes to us. The picture in Figure 10-21 is a horizontal system where the system in Figure 10-22 is a vertical system. Figure 10-21 is a carousel where Figure 10-22 would be a ferris wheel.

Figure 10-20 Maintenance Tool Crib

Courtesy of Global Equipment Co.

Figure 10-21 Carousel Conveyor

Courtesy of S.I. Handling Systems, Inc.

Figure 10-22 Vertical Carousel Storage and Retrieval System

Courtesy of S.I. Handling Systems, Inc.

Systems Required for Stores Department

Locator system. The locator system was discussed in Chapter 7. Every location has an address and the warehouse person must know how to reach any address without taking time to think.

Kitting system. Kitting is the process of pulling together a 1,000 sets of parts (one day's supply) for tomorrows production (see Figure 10-23). This inventory is pulled from the storeroom stock and placed on pallets or carts to be moved to the assembly line for tomorrows work. Kitting needs space for holding the material and material handling equipment to move it out of the storeroom to production.

Figure 10-23 Carts For Kitting

SINGLE SIDED

DOUBLE SIDED

NOTE: 800 lb. models have 5″ x 1¼″ (2 rd., 2 swl.) rubber casters.

Folds to 6¼″

The kitting system is important since having a full day's inventory on the assembly line means that we will not run out. If something was missing in the warehouse, we could have 16 to 24 hours to resolve the problem.

Inventory control system. The inventory control system controls the storeroom. Maintaining a proper level of inventory is the function of inventory control. The storeroom is sized to maintain this inventory. The movement of material into and out of the storeroom must be reported and entered into the inventory control system.

FABRICATION

The fabrication department is the department that produces parts for the assembly and/or packout lines. This fabrication starts with raw materials and ends with finished parts. The material handling facilities include containers, workstation handling devices, and mobile equipment.

Shop Containers

Shop containers are used to move parts in unit loads (see Figure 10-24). We may chop up large sheets of steel or coils of steel into small parts. These parts are collected in bins or boxes made of cardboard, plastic, or steel and moved to the second operation (see Figure 10-25). Shop containers are often stacked on pallets moved to the next machine and placed on the next machine. Machines can be filled by using a device that holds shop containers at the correct angle and position. Shop containers are used over and over again. These containers must be durable, stackable, and portable.

Figure 10-24 Shop Boxes in Fabrication Area

Courtesy of Streator Dependable Mfg.

Figure 10-25 Shop Containers

ALL PURPOSE **SHOP BOX**
Unique 3-way design!

NEST

STRUCTURAL TOTE

STACK

STEEL BOX

Courtesy of Global Equipment Co.

Tubs and Baskets

Tubs and baskets are larger shop containers (see Figure 10-26). Regular tubs and baskets are 4′ × 4′ × 42″ high. These are large sized containers. Parts on the bottom are often hard (and time consuming) to retrieve. For this reason, several special tubs and baskets are available.

Drop bottom tubs. A special rack holds the tub over an inclined slide (see Figure 10-27). Once on the rack, the bottom front is dropped and parts flow out the bottom of the tub which has been elevated to a good work height for the operator.

Drop side tubs or baskets. Not as good as drop bottom, but cheaper (see Figure 10-28).

Tilt stands. Tilt stands hold regular tubs and baskets on an angle to make retreiving parts easier (see Figures 10-29 and 10-30).

"V" stands. "V" stands are stands for smaller cartons and holds these cartons at work level on a 45° angle to allow for easy access to parts (see Figure 10-31).

Figure 10-26 Tubs and Baskets

Figure 10-27 Drop Bottom Tub

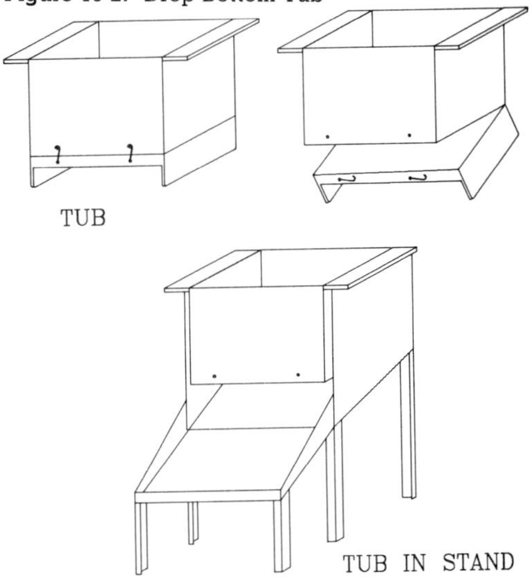

■ *HOLD N' FOLD*
■ *HEAVY DUTY RIGID*
■ *CORRUGATED*
■ *WORKINGTAINER*

TUB

TUB IN STAND

Courtesy of Steel King Industrial Containers.

Figure 10-28 Drop Side Basket

Courtesy of Steel King Industrial Containers.

FABRICATION

Figure 10-29 Tilt Tables and Stands

Table raises to any position and then tilts to place load in easy reach of the worker.

The retaining plate adjusts to different positions to place work at easily accessible levels by user.

Courtesy of Air Technical Industries.

Figure 10-30 Tilt Stand

Courtesy of Streator Dependable Mfg.

CHAP. 10: MATERIAL HANDLING EQUIPMENT

Figure 10-31 "V" Stand

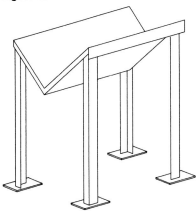

Scissor lifts or hydraulic lifts. A scissor lift will lift up a pallet of material to keep the material at a comfortable height (see Figure 10-32).

Dump hoppers. Dump hoppers can make the handling of material at a workstation almost effortless (see Figure 10-33). Dump hoppers will clamp a tub or basket of parts into position, lift that tub and tilt it up to 120°, spilling the parts onto a slide that can bring the parts up to the point needed. It is very efficient.

Figure 10-32 Scissor Lifts

Courtesy of W.W. Monroe Equipment Co.

Figure 10-33 Dump Hoppers

Dump Position   **Load Position**

Courtesy of Wilde Mfg., Inc.

Workstation Material Handling Devices

Counterbalances. Counterbalances hold tools above where they are needed and nearly eliminate the weight of the tool (see Figures 10-34a and 10-34b). They are one material handling device that takes the physical labor out of work.

Figure 10-34a Counter Balancers

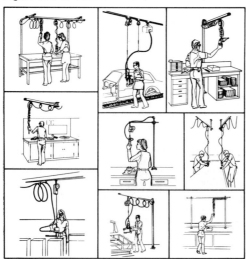

Courtesy of Aero-Motive Mfg. Co.

Figure 10-34b Counter Balancer

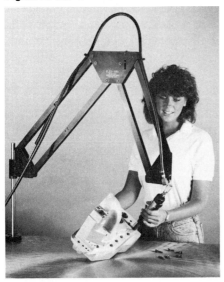

Courtesy of Flex Arm.

Vibratory feeders. Vibratory feeders orient, feed, count, and present a part to the next operator (see Figure 10-35). Many machines have parts feeders loading parts automatically. A toy maker needed 4 million little tires put on hubs. They set up two vibratory feeders (one for wheels and one for hubs) and two round wheels picked up a part each from the feeders and pressed the tire onto the hub automatically. A swingset manufacturer set up 20 vibratory feeders to assemble parts bags.

Figure 10-35 Vibratory Parts Handling System

Picture Taken by Author.

FABRICATION

Each feeder was connected to a control panel and the number of required parts entered into the controller. All 20 feeders counted out their parts and stopped. When all 20 feeders had stopped, hoppers were opened and the parts fell into a bucket on the conveyor. The conveyor advanced one feeder and the feeders started again. The bags were formed, packed, and weight checked all automatically.

Rivets, eyelets, screws, and bolts are fed into machines that use these fasteners by means of vibratory feeders.

Waste disposal. Removing waste from workstations requires special material handling equipment (see Figure 10-36). Dump hoppers are like that pictured in Figure 10-33. Chip removal from cutting machines remove the cutting oils and place the chips in a dump hopper. Trash compactors reduce waste removal costs, and paper bailers will turn trash costs into profits. Waste disposal is a big job where material handling equipment can improve performance and reduce costs.

Walking beams. Walking beams can continually load and unload machines eliminating the need for an operator doing any material handling (see Figure 10-37). Walking beams pick up a part, move the part into the machine, lower the part, and return to the starting point. Two walking beams work on the same machine—one loading and one unloading.

Figure 10-36 Waste Disposal

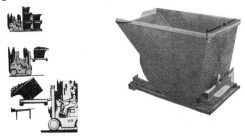

Courtesy of Global Equipment Co.

Figure 10-37 Walking Beam Load and Unload Press

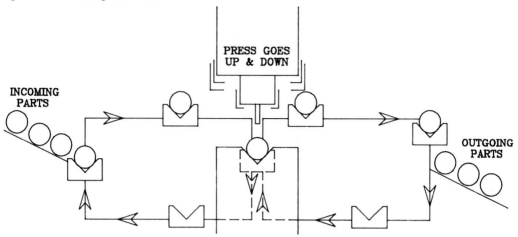

Ball tables. Ball tables have ball bearings mounted on a table top to allow the easy movement of heavy material (see Figure 10-38). A 200-pound sheet can be moved with only 10 pounds of force.

Powered round tables. A workstation can be built on a round table and indexed automatically (see Figure 10-39). Golf club irons have the need for a 2″ hole drilled in the hozzle (shaft hole). The hole is tapered, tapped, and the top is spot faced. All these operations were done at the same time on a round table.

Figure 10-38 Ball Table

Ball Transfer Table

Figure 10-39 Power Round Table

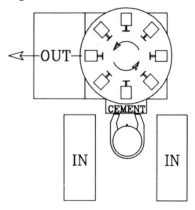

Courtesy of Hytrol Conveyor Co.

FABRICATION

Figure 10-40 Jib Crane

Courtesy of Air Technical Industries.

Jib cranes. Jib cranes are a lifting device attached to a boom (see Figure 10-40). The boom is mounted to the top of a mast (upright beams). The mast rotates 360° around the boom. Heavy parts or tools can be loaded into machines. A 20′ boom mounted between four machines can service all four machines.

Vacuum or magnetic lifts. Mounted to a jib crane or in large plants, bridge cranes, a vacuum, or magnetic lift hoists long heavy sheets of material (see Figure 10-41). Skins for aircraft are moved with vacuum lifts.

Figure 10-41 Manipulator

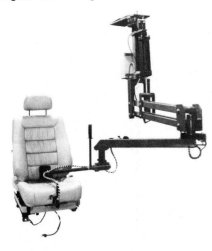

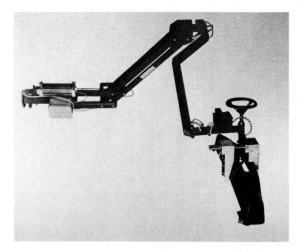

Courtesy of TDA Buddy Systems.

Figure 10-42 Welding Robot

Courtesy of Miller Electronic Mfg. Co.

Robots. Robots can be used to load and unload machines (see Figure 10-42). They can also be used to weld and paint.

Mobile Fabrication Equipment

Moving material around the fabrication area requires equipment that can follow various paths. These paths can change with each different manufactured part's requirements. The equipment covered in this section is not in any pre-set order and can be easily rearranged.

Slides and chutes. Slides and chutes are as a simple as a child's playground slide (see Figure 10-43). Material is placed on the top of the slide by the operator who has just finished their operation. The part slides down to the next operator by use of gravity. Slides and chutes can be made of wood, plastic or steel. They can be easily moved.

Skatewheel and roller conveyor (non-powered). Skatewheel and roller conveyors come in 10′ sections and can be combined to make any length (see Figure 10-44). They can be easily moved for a change of direction, and the slope can be adjusted to make the parts roll. If some parts won't roll, they can be put in shop containers or on wooden boards in order to make them roll.

Figure 10-43a Robot Dimensions and Work Envelope

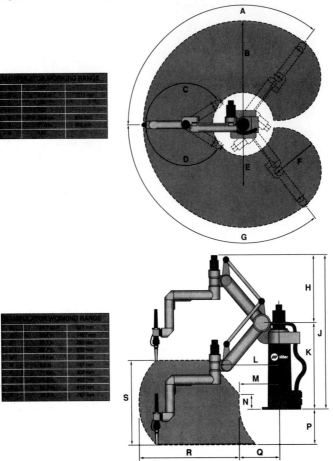

Courtesy of Miller Electronic Mfg. Co.

Skatewheel and roller conveyors are very flexible in that they can be made to follow any path. "V" stands (Figure 10-31) have been combined with skatewheel rollers to create a material handling system that connects two workstations and automatically feeds parts to a convenient position for the operators to place and pick up parts without walking or bending.

Lift conveyors. Lift conveyors can move parts from near the floor to any elevated position (see Figure 10-45). They are used on automatic machines where the part drops out of the machine on a chute to the bottom of the lift conveyor and lifted to the top of a tub where hundreds or thousands are collected. Lift conveyors (sometimes called *bucket conveyors*), can move water, grain, coal, or just about anything where a lot of volume is needed.

Figure 10-43b The Chute and Slide

■ The Chute

The Chute is an easy, fast, and economical way to convey products from one level to a lower level. There are no moving parts to wear out and the Chute can be quickly installed almost anywhere.

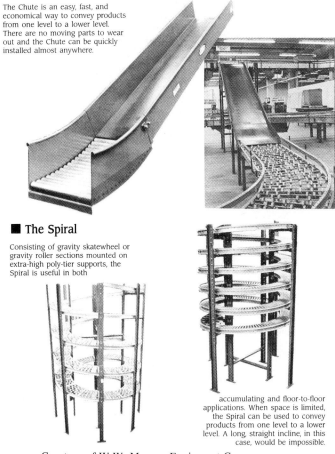

■ The Spiral

Consisting of gravity skatewheel or gravity roller sections mounted on extra-high poly-tier supports, the Spiral is useful in both

accumulating and floor-to-floor applications. When space is limited, the Spiral can be used to convey products from one level to a lower level. A long, straight incline, in this case, would be impossible.

Courtesy of W.W. Monroe Equipment Co.

Auger conveyor. Auger conveyors are tubes with a screw inside (see Figure 10-46). The turning of the screw pulls and pushes the material in the direction of the screw rotation. Grain and wood chips are moved this way.

Vibratory conveyor. Vibratory conveyors move parts down a chute or slide by vibration. Inclined vibratory conveyors are used in the separation of parts such as sand in casting or parts from tumbling media like plastic pallets, corn cobs, and rocks (see Figure 10-47).

Monorail trolley. A monorail tolley is a single rail over the workstation or between two workstations which can move parts of tools along a fixed path (see Figure 10-48). If a heavy tool is needed anywhere along a 20′ path, place a monorail above this path and hang the tool from the rail.

Figure 10-44 Skatewheel and Roller Conveyor

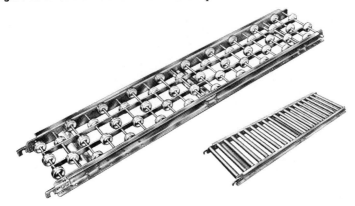

■ Six-Way Skatewheel Switch

This special "Six-Way" Skatewheel Switch is designed to transfer products from two diverging lines to a main line. This unit can also be provided with special removable guard rails.

Courtesy of Hytrol Conveyor Co.

Figure 10-45 Lift Conveyor

Courtesy of Hytrol Conveyor Co.

Figure 10-46 Auger or Screw Conveyor

SCREW CONVEYORS

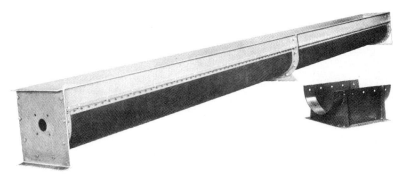

Courtesy of Conveyor and Drive Equipment Co., Inc.

Figure 10-47 Vibrating Equipment

SYNTRON VIBRATORY FEEDERS

With flat bottom trough   With tubular trough

Courtesy of Conveyor and Drive Equipment Co., Inc.

Figure 10-48 Monorail Trolley Conveyor

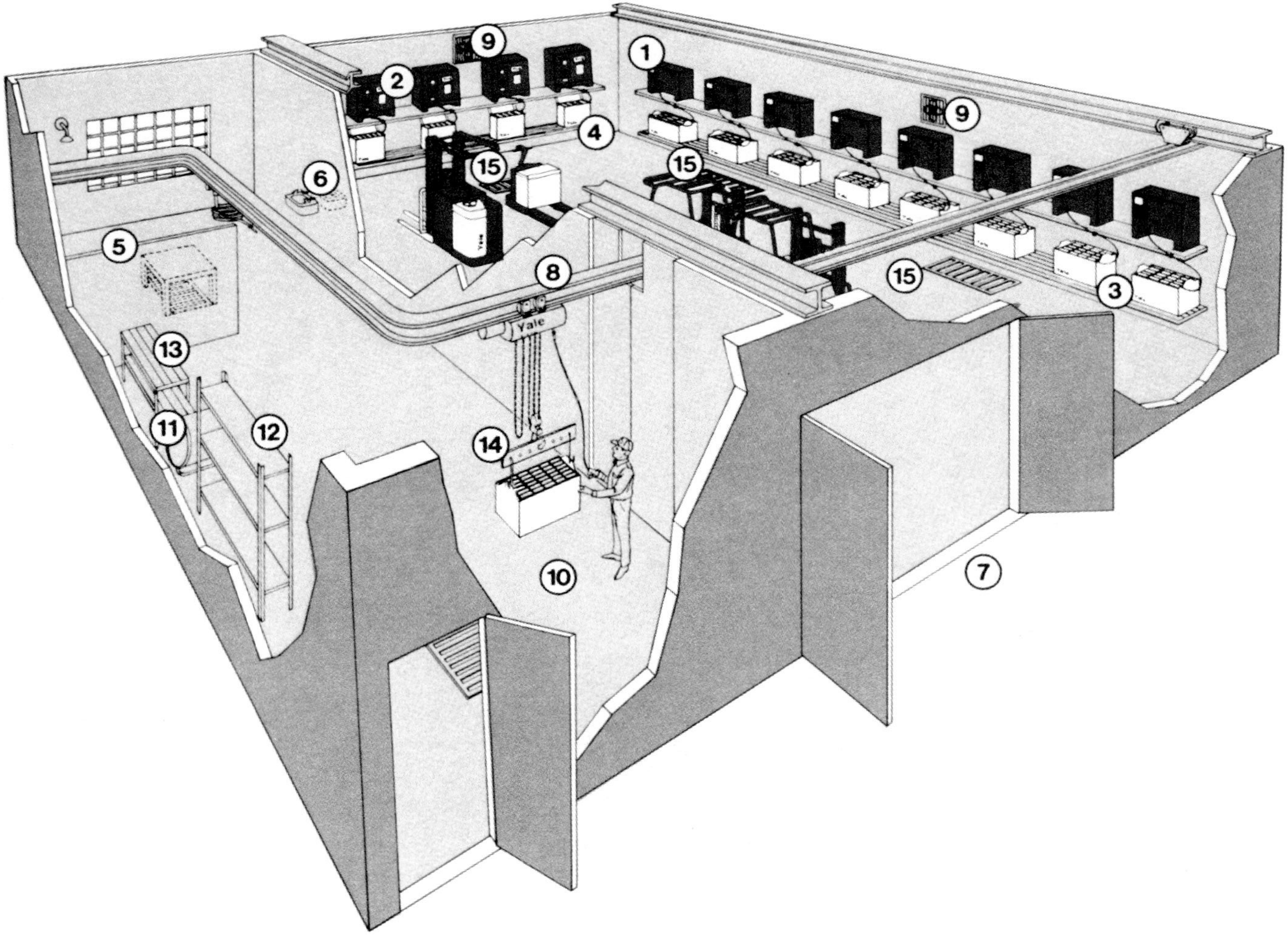

Courtesy of Yale Materials Handling Corp.

214

Powered hand trucks. Powered hand trucks are just like hand trucks (see Figure 10-5b) except they have a battery attached (see Figures 10-49 and 10-50). They can lift and move greater weights and are easier to control. For short distances (like within a department) they are more cost efficient than fork trucks.

Figure 10-49 Powered Hand Truck

Courtesy of Yale Materials Handling Corp.

Figure 10-50 Powered Hand Truck

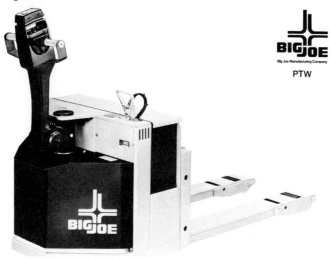

Courtesy of Big Joe Manufacturing Co.

FABRICATION

Many assembly operations, especially small assemblies and sub-assemblies are just like fabrication workstations and will use the equipment just talked about in the previous section (fabrication). Counterbalances (Figure 10-34), vibratory feeders (Figure 10-35), tilt stands (Figure 10-29), dump hoppers (Figure 10-33), shop containers (Figure 10-26), and tubs and baskets (Figure 10-27) all are used in assembly, but when we speak of assembly material handling equipment, almost everyone thinks of conveyors. There are many different conveyors. This section will cover the most popular.

Belt Conveyors

Belt conveyors are endless loops of fabric that can be any width by any length (see Figure 10-51). Belt conveyors eliminate the need for assemblers to move assemblies into and out of their workstation. They also eliminate the need to hold the base unit. Conveyor belt speed and work height should be adjustable. Stops can be built over the belt to deliver assemblies to a workstation and hold them until the task is complete. Belt material can be cloth or rubber. The belt can run over sheet metal or rollers.

Figure 10-51 Belt Conveyor

ROLLER BED
BELT CONVEYOR

For moving heavier "units" loads
from department to department, for
assembly, inspection, or packaging.
Roller bed reduces belt friction for
greater capacity.

★*Moves Heavier Loads*

★*Reversible (with Center Drive)*

★*Heavy-Duty Frame*

★*7 Belt Widths—12 in. to 36 in.*

★*Sealed Bearings*

★*Adjustable Floor Supports*

LIGHT-DUTY
BELT CONVEYOR

Easily set up, work table type for
assembly line operation, inspection,
sorting, and packing.

★*Economical*

★*Reversible (with Center Drive)*

★*Floor or Ceiling Supported*

★*Smooth Slim Bed*

★*10 Belt Widths—6 in. to 30 in.*

WIRE MESH
BELT CONVEYOR

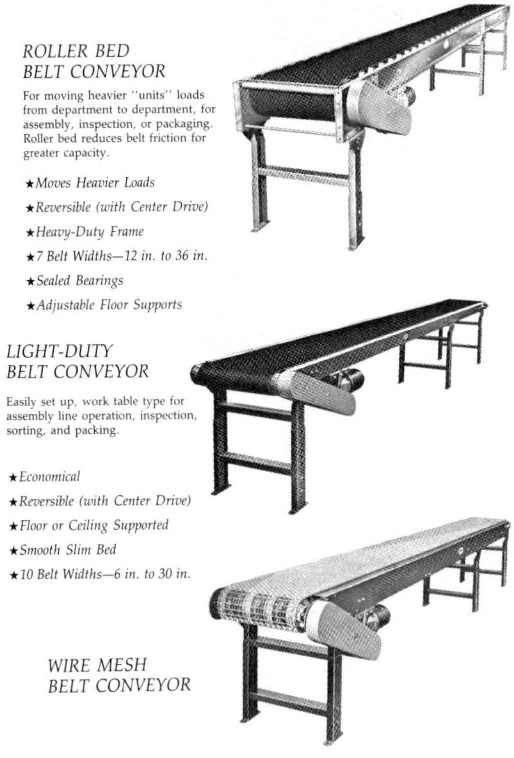

Courtesy of Hytrol Conveyor Co.

CHAP. 10: MATERIAL HANDLING EQUIPMENT

Figure 10-52 Roller Type Conveyors

LIVE ROLLER CONVEYOR

The Model "190-SP" live roller "spool" conveyor is a general transport conveyor with the capabilities of accumulating products with minimum back pressure. Quiet operation, versatile design, and easy installation are standard features that make the "190-SP" conveyor a valuable component in operations requiring high performance with minimal downtime.

★*12 Bed Widths*

★*Minimum Back Pressure*

★*Single Drive Powers Curves—Spurs—Straights*

★*High Speed Capabilities*

CHAIN DRIVEN LIVE ROLLER CONVEYOR (ROLL-TO-ROLL)

The heavy, rugged design of the "25-CRR" & "26-CRR" allows it to be used for conveying higher load capacities such as loaded pallets and drums. Chain driven rollers make it ideal for wash-down operations and conveying oily parts in bottling and steel industries, foundries, etc.

★*Center Drive*

★*12 Bed Widths—22-1/4 in. to 54-1/4 in.*

★*Removable Sealed Bearings*

★*Reversible*

★*Adjustable Floor Supports*

Courtesy of Hytrol Conveyor Co.

Powered Roller Conveyor

The powered roller conveyor (see Figure 10-52) performs just like a belt conveyor and looks like a non-powered roller conveyor like that shown in Figure 10-44. Moving boxes over a fixed path for long distances is a good use of the powered roller conveyor.

Car Type Conveyor

A car type conveyor can be made by attaching fixtures onto a cable and pulling the cable around a fixed path (see Figure 10-53). The car type conveyor looks like a small railroad train with the flat cars totally filling the looped track. Think of a child's train set and visualize the track as being totally full of flat cars. Instead of pulling these cars by an engine, we will set up a cable system to pull these cars along at a uniform speed. On top of these flat cars, we can build holding fixtures to hold any shape.

Figure 10-53a Car Type Conveyor

Courtesy of Webb-Stiles Company.

Figure 10-53b Car Type Conveyor

Courtesy of Webb-Stiles Company.

Slat Conveyor

Slat conveyors are narrow slates of wood or metal attached to chains (see Figure 10-54). The slats will travel down a parallel pair of chains to the end of the line and run back to the beginning under the line just like belt conveyors. The lumber industry cuts lumber and lets it drop on a slat conveyor made out of 2" × 6" × 20' long slats. This belt runs 200' away from the saw, and along both sides, laborers separate the sizes and grades of lumber and place lumber on carts.

In drink bottling plants, bottles or cans are carried through the filling, capping, and labeling machines by slat conveyors made of thin 6" × 4" metal slats. TV sets are assembled on a slat conveyor made up of 2" × 4" × 4' long with electrical plug-ins every 2'. Caterpillar Tractor Company assembles their largest tractors on a slat conveyor built at floor level. The conveyor only moves a few feet per hour, but material can be moved on and off the slat conveyor with fork trucks. People walk on and off the conveyor without even knowing they were on a conveyor. The caterpillar slat conveyor is made of steel slats about 1/2" thick, 12" wide, and 20' long.

Figure 10-54 Slat Conveyor

Courtesy of Hytrol Conveyor Co.

Tow Conveyor

Tow conveyors pull carts around a fixed path (see Figure 10-55). The power can be overhead or under the floor, but both do the same job. One advantage of a tow assembly line is that one unit can be removed from the line without stopping the line. The tow conveyor carts can carry a wide variety of products. The fixtures mounted to the carts will be for a specific product.

Overhead Trolley Conveyor

Overhead trolley conveyors can go anywhere (see Figure 10-56). One conveyor manufacturer uses a slogan like this: "Anywhere and Everywhere With Unibilt." Overhead trolley systems can carry parts through heat treat, washing, painting, drying to the assembly department. They can be loaded and unloaded at floor

Figure 10-55a Towline

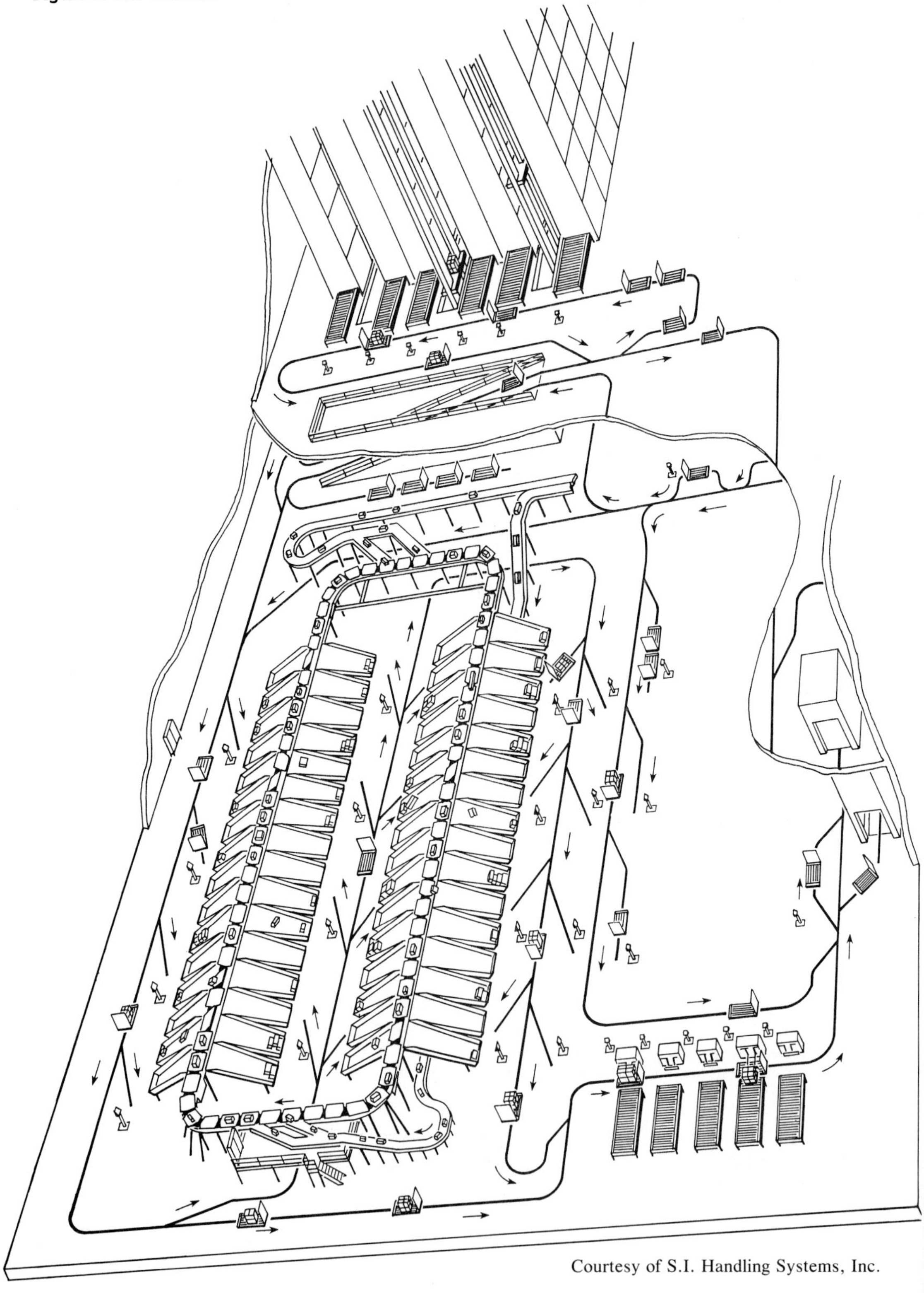

Figure 10-55b Towline Conveyor

Courtesy of S.I. Handling Systems, Inc.

Figure 10-55c Towline Conveyor

ASSEMBLY AND PAINT

Figure 10-56a Overhead Trolley Conveyor

Courtesy of Richards-Wilcox.

Figure 10-56b Trolley Conveyors

Courtesy of the Buschman Company.

CHAP. 10: MATERIAL HANDLING EQUIPMENT

level, and then raise to the ceiling for traveling over the plant's equipment and people. Trolley conveyors are cable or chain pulled through channels of I beams with a single drive unit. Below the trollies are mounted hooks and racks for carrying the parts. This study, alone, could be a career.

A simple S hook is the most common method of hanging parts from the trolley, but the hooking systems can get quite complicated. Parts can be spun on the overhead trolley conveyor by placing a wheel on the hook and a stationary rub bar mounted along the area where you want the part to spin. Parts can be turned by placing an "X" on top of the hook, and a stationary pin for every 90° of turn desired.

Power and Free Conveyor

Power and free conveyors are dual track trolley conveyors with one track for the power line and the other track for carrying the trollies (see Figure 10-57). The advantage is that a single part can be stopped without stopping the line. If you are pouring molten iron into mold, you would not want the mold moving, so we stop it long enough to pour, then reconnect it to the power line. Power and free can divert product no different lines or place in hold areas.

Figure 10-57 Power and Free Conveyor

Courtesy of Richards-Wilcox.

PACKOUT

Packout is usually the end of assembly and many of the same material handling devices are used here. Packout is usually packaging one unit for shipment. But, sometimes, packaging will include putting many products into one package. Material handling equipment has improved the quality and efficiency and packaging. The following equipment are some of the many pieces of equipment used in the packout department.

Box Formers

Box forming can be accomplished automatically and wrapped around the product being packaged (see Figure 10-58). Soft drink bottling plants are a good example of this.

Automatic Taping, Gluing, and Stapling

Closing boxes and sealing them can be accomplished automatically on the packout conveyor (see Figure 10-59).

Figure 10-58 Box Former

Palletizers

After the boxes are filled and closed, they are automatically stacked on pallets per a prearranged program and the full pallets are moved out to a pickoff area where a truck moves them to the warehouse (see Figure 10-60).

Pick and Place Robots

Pick and place robots also make up pallets of finished products (see Figure 10-61). The robot can handle a number of different packages at the same time, but the function is the same as a palletizer.

CHAP. 10: MATERIAL HANDLING EQUIPMENT

Figure 10-59 Taping

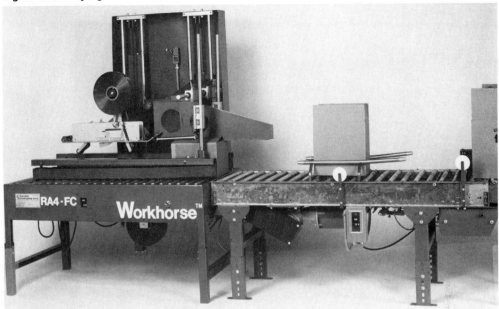

Courtesy of Durable Packaging Corp., 3139 West Chicago Avenue, Chicago, IL.

Figure 10-60 Robotic Palletizer

Courtesy of GM Fanuc Robotics Corporation.

PACKOUT

Figure 10-61 Pick and Place Robot

Courtesy of GM Fanuc Robotics Corporation.

226

Banding

Banding boxes closed can be accomplished automatically by placing a bander around the packout conveyor (see Figure 10-62). Banding will also hold many packages on a pallet. Banding is used when packages cannot hold themselves on pallets. When a carton is nearly square, they will not tie (hold themselves on the pallet) so banding is needed.

Stretch Wrap

Stretch wrap is like banding in that it holds packages together on a pallet (see Figure 10-63).

Figure 10-62 Banding

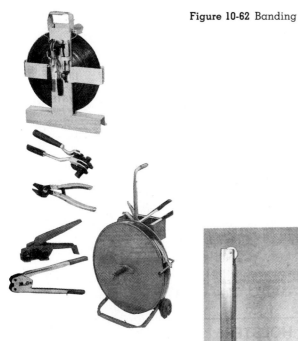

Figure 10-63 Stretch Wrap

STRETCH-FILM
DISPENSER

Courtesy of Global Equipment Co.

Courtesy of Global Equipment Co.

Warehousing looks similar to stores because shelves, racks, pallets, and some trucks are similar. These pieces of equipment will not be rehashed here. Some unique equipment for warehousing will be the thrust of this section. The functions of a warehouse is to pick customers orders and prepare them for shipping. The first group of equipment to be discussed will be used for picking customer orders.

Picking Carts

Customer's orders can be picked from shelves and placed on picking carts. Tool, drug, audio tape, and book warehouses would use this kind of equipment (see Figures 10-64 and 10-65).

Figure 10-64 Picking Cart

Figure 10-65 Picking Truck

Courtesy of Global Equipment Co.

Courtesy of Crown Equipment Corp.

Figure 10-66a Flow Rack

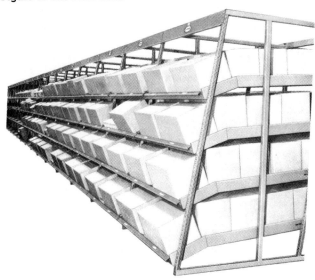

Courtesy of Hytrol Conveyor Co.

Gravity Flow Bins

When the product is small, and high volume is sold, many parts can be stored in a small area reducing the pickers need to travel great distances (see Figure 10-66). Drug warehouses use this system for their "A" items.

Tractor-Trailer Picking Cart

When picking larger orders, like grocery store warehouses, an order picker would drive a tractor pulling many trailers (see Figure 10-67). Remote control tractors are often used so the order picker can be retrieving groceries for a rear trailer and still move the tractor.

Trailers that follow one another (tracking) are very important. There are two techniques used to get trailers to track well:

1. The front wheels turn one way while the back wheels turn the opposite way.
2. Two load bearing wheels are placed in the middle of the trailer, and pilot wheels are mounted in the middle of the front and rear. The pilot wheels just keep the trailer level.

Once the trailers have been filled, the driver takes them to shipping to be loaded on an over-the-road trailer.

WAREHOUSING 229

Figure 10-66b Automated Warehouse

The Automated Warehouse... Integrated SI Order Selection Systems

SI Handling Systems, Inc. can provide an order selection system for your needs. Each SI system is designed to provide the most efficient picking process for specific types of product, orders and fulfillment needs. Yet each system can be fully integrated into an overall warehouse control system working together with other picking systems for optimized efficiency.

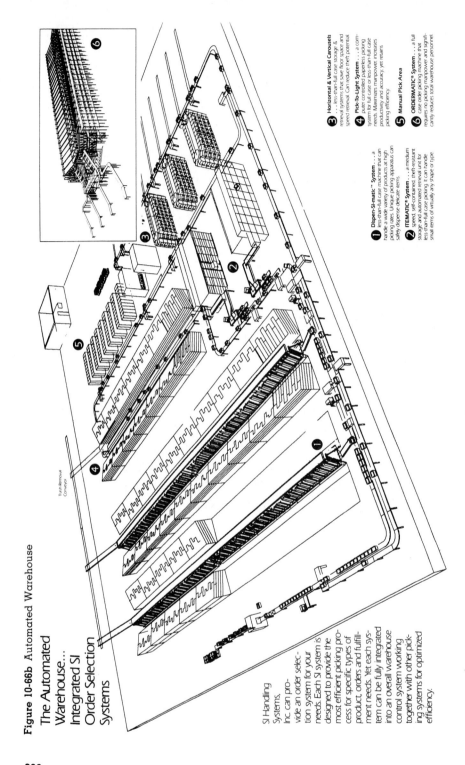

Turn Removal Conveyor

❶ Dispen-SI-matic™ System . . . a less-than-full case machine that can handle a wide variety of products at high picking rates. Unique picking apparatus can safely dispense delicate items.

❷ ITEMATIC® System . . . a medium speed, self-contained, theft-resistant storage and automated retrieval unit for less-than-full case picking. It can handle small items of virtually any shape or type

❸ Horizontal & Vertical Carousels . . . less-than-full case storage and retrieval systems that save floor space and speed retrieval. Can reduce theft potential.

❹ Pick-To-Light System . . . a computer controlled paperless picking system for full case or less-than-full case needs. Maximizes manpower productivity and accuracy, yet retains picking efficiency.

❺ Manual Pick Area

❻ ORDERMATIC® System . . . a full case order picking machine that requires no picking manpower and significantly reduces total warehouse personnel

Figure 10-67 Tractor/Trailer Picking

Courtesy of Crown Equipment Corp.

Clamp Trucks

Clamp trucks are a special fork truck which eliminated the use of pallets (see Figure 10-68). Two 4 × 4 foot plates squeeze the stacked boxes together, lifts them and moves the material into and out of the warehouse. Toy industries, gas grills, appliances, tackle boxes, and many more industries that produce large volumes of large items use this tool for space utilization and efficiency reasons.

Rotary Conveyor Bins

Rotary conveyor bins bring the product (small) to the picker saving all the walking. Spare parts warehouses use this technique (see Figure 10-69).

Figure 10-68 Clamp Truck

Courtesy of Crown Equipment Corp.

Figure 10-69 Rotary Conveyor Bins or Horizontal Carousel Conveyor

Courtesy of White Storage & Retrieval Systems, Inc.

CHAP. 10: MATERIAL HANDLING EQUIPMENT

Vertical Warehouse and Picking Cars

The vertical warehouse could have 40 shelves high which are 300' long on both sides of a 4' aisle (see Figure 10-70). Eight thousand shelves would be available for 8,000 different items. The order picker picks up all the orders in the warehouse for this group of products (they were placed in picking order first) and proceeded to the first location. The order picker rides on a pallet that goes up and down with the stock as picked. One pass through the aisle, the picker unloads at shipping and returns for new orders. The cart does not need to be steered, it is on rails or has wheel guides. Large catalog distribution centers use this technique. (See Figure 10-71.)

Packing Station

Once all the order has been picked, it must be packed for shipping (see Figure 10-72). Small parts must be packaged properly into larger cartons and wrapped so not to be damaged in shipment. Larger products may just be addressed, some may be full truck loads, but some preparation is always needed, so a packaging station is required (see Figure 10-72). A weight scale built into the packaging station (see Figure 10-13) is many times, very desirable.

Figure 10-70 Vertical Warehouse Picking   **Figure 10-71** Vertical Warehousing

Courtesy of S.I. Handling Systems, Inc.   Courtesy of Yale Materials Handling Corp.

Figure 10-72 Packaging Work Station

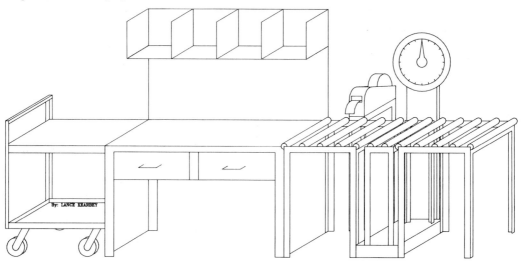

Shipping Containers

Most shipping containers are pallets or the packout carton, but sometimes it may be the size of a tractor trailer (see Figure 10-73). These are called *cargo containers* and can be shipped over the road, on railroad flat cars and on ocean freighters. These containers can be sealed by the shipper and next opened until received by the customer.

Figure 10-73 Shipping Containers

Courtesy of Steel King Industries, Inc.

Figure 10-73 (cont'd) Shipping Containers

BULK MATERIAL HANDLING

Bulk material handling is a very special subject. It deserves much more attention than will be given in this book due to our limited space for this discussion. Bulk material means a lot of material (for example, coal in a coal mine through the power plant; lumber and paper products from the forest through the mills and plants; ore from the ground through the mills; oil from the ground to the service station; and grain from farms through mills and plants). The one advantage these bulk mills and plants have is that one (or a few) materials make up their materials list, and we can concentrate on this one item. Bulk material handling equipment varies in size from a pump for an oil plant to a conveyor systems several miles long. The following list of bulk material handling equipment is grossly understated, but if you join one of these industries, the list of equipment will be specific to that industry, and you will become familiar with that group of equipment very fast.

Bulk Material Conveyors

Troughed belt conveyor. Belt troughed conveyors are concave and look like a long feed trough (see Figure 10-74). Coal industry use these conveyors to move coal from the face of the mine to the elevators and from the top of elevators to the coal pile. Logging industries use troughed conveyors for moving logs from rivers or lakes to the mill ponds. A side light is that mill ponds, which float logs to the mills, are material handling devices.

Screw conveyor. This type of equipment (also called augers) was discussed in the fabrication area, but much more use is made of screw conveyors in processing plants like paper mills, bakeries, and feed mills (see Figure 10-46).

Figure 10-74 Troughed Bed Belt Conveyor (Roller Bed)

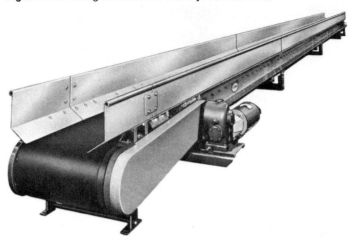

Courtesy of Hytrol Conveyor Co.

Vacuum delivery systems. A vacuum system is a system of tubes moving pellets or powders from tank cars to storage towers to equipment (see Figure 10-75). Plastics is a good example of this system. A vacuum system makes the material handling labor free. The storage towers (or silo) are also material handling devices.

Figure 10-75 Vacuum Delivery System

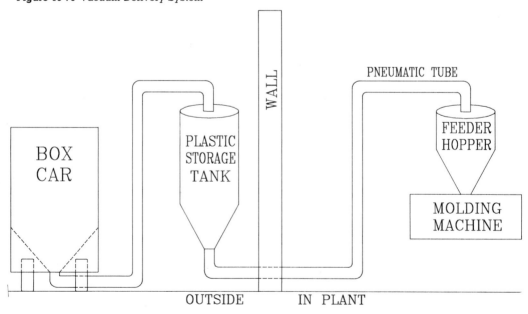

CHAP. 10: MATERIAL HANDLING EQUIPMENT

Figure 10-76 Pumps and Tanks

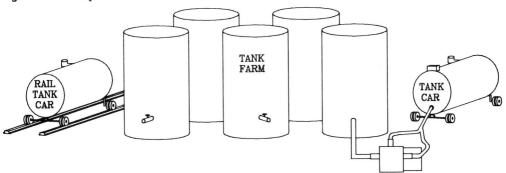

Pumps and tanks. Oil, drinks, most liquids and semi-liquids are moved from tankers to tanks to filling stations by pumps (see Figure 10-76). Pumps have hoses leading into and out of them, flow meters to measure volume, gauges to measure tank fullness (tanks are a part of material handling) and petcocks to tap off samples of the product for quality control.

Conveyor systems. If your bulk products are cartons, a system of diverted conveyors may be used (Figure 10-77). UPS, Sears Distribution, J.C. Penney's distribution use major conveyor systems for the distribution of many packages.

Figure 10-77 Bulk Carton Handling

BULK MATERIAL HANDLING

Figure 10-77 (cont'd) Bulk Carton Handling

COMPUTER-INTEGRATED MATERIAL HANDLING SYSTEMS

State-of-the-art material handling systems, smart systems, and world class material handling systems all communicate the need to keep improving our cost performance everyday. We are in worldwide competition and material handling costs are a major component of product cost, so we need to keep improving. The technology is here to eliminate large pieces of product costs. Material handling equipment is as important as any machine making parts and the modern technology of material handling equipment has been keeping pace with all other equipment. One piece of equipment (actually a whole system) is the *automatic storage and retrieval systems* (ASRS). ASRS will automatically put product or parts away, or take product out, move it to where required and adjust the inventory level at both ends of the move (see Figures 10-78 through 10-81). ASRS systems are typically very tall (60 feet and over) and very large areas. ASRS is made up of:

1. racks;
2. shuttle cars;
3. bridge cranes;
4. computer control center; and
5. conveyor systems.

CHAP. 10: MATERIAL HANDLING EQUIPMENT

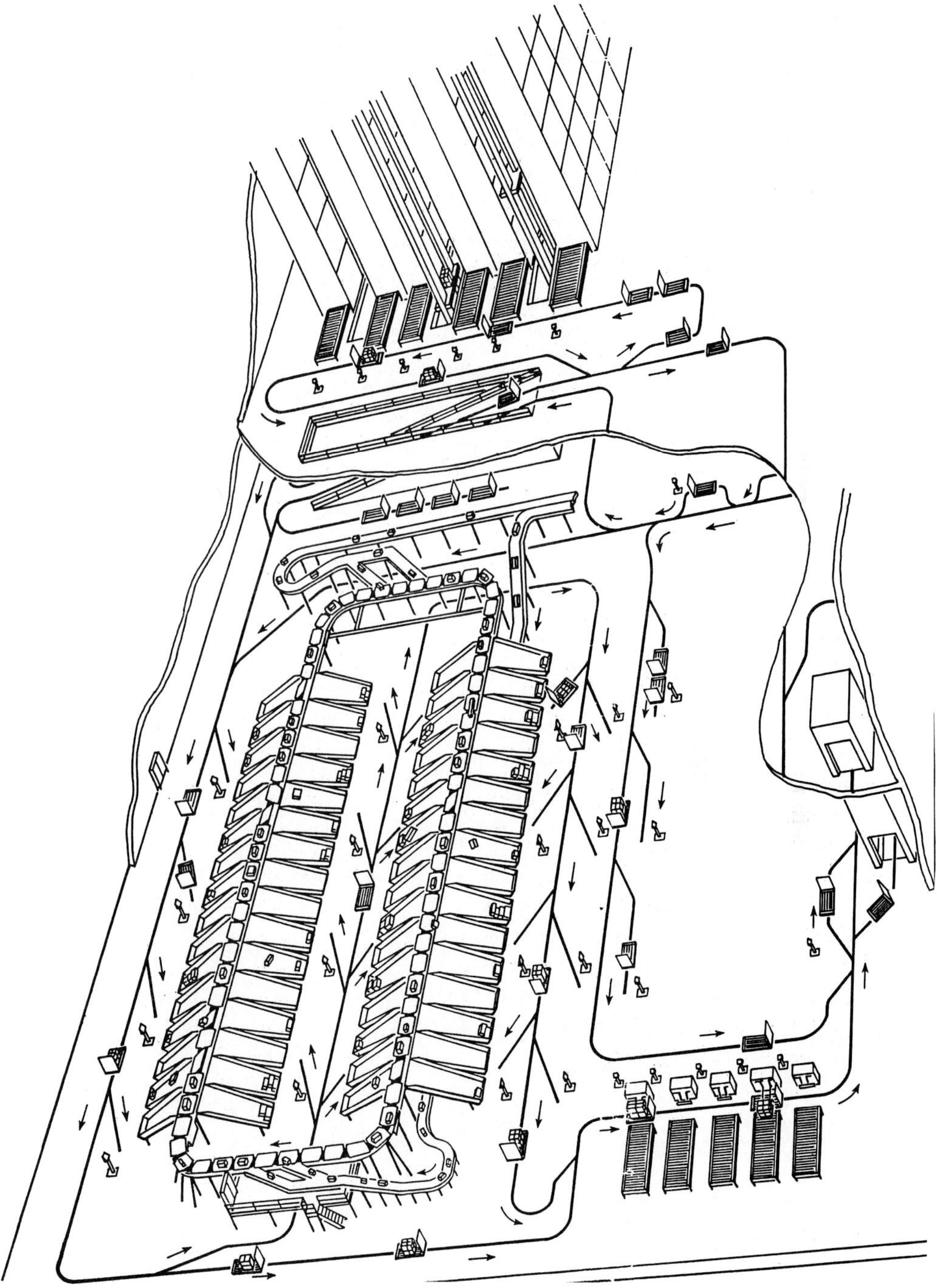

Figure 10-78 ASRS

Figure 10-79 ASRS

Figure 10-80α Automatic Storage and Retrieval Systems

Courtesy of Erman Incorporated.

CHAP. 10: MATERIAL HANDLING EQUIPMENT

Figure 10-80b Automatic Storage and Retrieval Systems

Figure 10-81 ASRS Building Construction

Courtesy of S.I. Handling Systems, Inc.

COMPUTER-INTEGRATED MATERIAL HANDLING SYSTEMS

Chapter 11

Office Layout Techniques and Space Requirements

The office layout process is very similar to the manufacturing plant layout procedure. Many of the techniques used to study the flow of material are used to study the flow of paper, information, and people in an office. The activity relationship diagram, worksheet, and dimensionless block diagram studied in Chapter 5 are even more useful in office layouts because office sizes are closer to the same size than are manufacturing plant departments. The collection and the analysis of data as discussed in Chapters 2 and 3 will be an important part of office layout. Instead of studying material flow, we'll study information and paperwork flow. An office layout technologist must learn and understand office systems and procedures in order to establish proper placement of offices. We will study a systems and procedure analysis technique.

Who works in the office, what tasks are performed in the office, how people are organized into departments, and how these departments relate to each other are all extremely important questions to keep in mind when creating an office layout. An *organization chart* is an informative tool used to communicate the relationships among the departments and their people.

GOALS OF OFFICE LAYOUT DESIGN

The goals of office layout design will help the designer keep on tract and will give the technologist a way of evaluating the many alternatives. Some of the most common goals are:

1. Minimize *project cost*. Cost consciousness is important. The layout technologist must be responsible for recommending facilities that are cost effective. Buying the cheapest desk may not be cost effective if we need to replace it soon. There is value to good-looking facilities for customer and employee morale and attitude. Cheaper furniture seems to have harder surfaces which add to noise. Being cost conscious means that we want our money's worth and we are willing to shop around for the best facilities for our money.

2. The *productivity* of our employees is important. We do not want them walking long distances, performing useless work, using slow equipment, all of which can make life unpleasant. We want to promote an effective use of our people.

3. Office layouts must be *flexible*. One thing for sure is that office layouts will change. We must have the ability to expand or shrink overnight. We will talk about specific flexible furniture later in this chapter.

4. *Cleaning* and *maintaining* office space is costly. The type of layout and the equipment we buy will affect this cost.

5. *Noise* must be kept to a minimum. The fabrics on the walls, floors, and ceilings affect the noise level.

6. *Material flow* (paper and supplies) as well as people flow distances must be held to a minimum. The further we walk or move material, the more the cost. Good flow analysis will minimize these distances.

7. Create a *pleasing atmosphere* in which to work in order to promote pride and productivity.

8. Minimize *visual distractions*. Panels and furniture can be used to provide at least semi-private offices.

9. Create a *pleasing reception area*. First impressions or opinions of our company are produced in the visitor's reception area. Are we organized, efficient, and neat, or are we sloppy and messy?

10. *Energy costs* can be affected by the layout and must be minimized wherever possible. Windows, full walls, doors, and the like will all affect the energy costs.

11. Each employee needs adequate *work space* and *equipment*. Office layouts must address every office worker's needs.

12. Provide for the *convenience* of employees. Restrooms, lockers (or coat racks), lunch rooms, and lounges must be conveniently located to prevent long trips away from their offices.

13. Provide for the *safety* of employees. Aisle sizes, stairways, machines, and clutter can cause safety problems. Our layout plan must consider the safety aspects of our office.

Office layouts vary in complexity from supervisor's standup desks located in the middle of a production department to an office complex housing hundreds of office employees. Office space costs more per square foot than does manufacturing or distribution space, so space use is very important. Costs per square foot exceed $25 per year in large cities and less than $5 per square foot per year in small towns. The advantage of being "uptown," "downtown," or in highly populated areas is being close to many other businesses, transportation, communications, and services. The disadvantages are congestion and costs. Many corporate offices are located in major business centers for convenience to other businesses, but their manufacturing plants and supporting offices are located in rural areas where space costs (and living costs) are usually less. Our discussions are focused on manufacturing plant office layouts instead of corporate offices, but the techniques and processes are the same.

Supervisor's Offices

Manufacturing plant supervisor's offices are good starting points for office discussions because they are small and a "feel" for space can be developed early. A 10' × 10' portable office located in the middle of a production department is shown in Figure 11-1. These portable offices can be moved as one large unit. Air conditioning units are attached to one wall because a closed area which is so small can soon get very uncomfortable. Shipping, receiving, maintenance as well as production supervisors could use this type of an office construction.

Supervisors should be located where they are immediately accessible to their employees. Having a line-of-sight view can improve communication. Supervisors also need to meet with employees in confidence at times and this type of office provides the necessary privacy. Discipline should always be carried out in private. If private offices are not available, conference rooms must be provided where private meetings can be held.

Some supervisors use standup desks located in the middle of their production area. Figure 11-2 shows such a facility. A stool may also be provided, but the stool should be high enough to allow the supervisor to either work standing up or sitting down.

Open Office Space

Open office space (see Figure 11-3) (also called *bullpens*) are large rooms which house many people. Open offices are very popular because of the following reasons:

1. *Communications* are easier. To talk with someone requires only the action of lifting the head and speaking. To know if someone across the room is available, just look.
2. Common *equipment* is accessible to more people.

Figure 11-1 Supervisor's Office

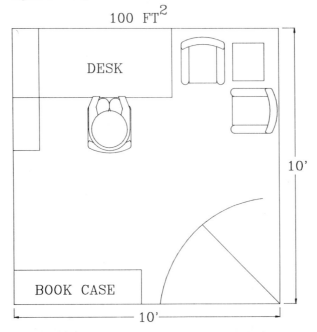

100 FT^2

DESK

10'

BOOK CASE

10'

8'H OFFICE 7'4" INSIDE
PRICE INCLUDES
ELECTRIC KIT

3" THICK SOUND
CONDITIONED PANELS

Courtesy of Global Equipment Co.

3. Less *space* is required, as compared to private offices.

4. *Heating, cooling* and *ventilation* costs and problems are minimized because one big room is easier to work with than the same area divided into private offices. Walls are the big enemy of good circulation. Open office construction eliminates walls.

5. *Supervision* of the people in an open office is easier. Doors and walls make supervision more difficult.

6. *Layout changes* are quicker and less costly in open offices. Moving desks around a large room is easier than negotiating aisles and doors.

Figure 11-2 Standup Desk

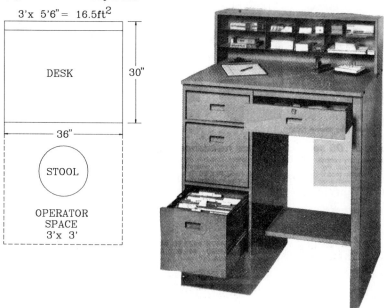

Courtesy of Global Equipment Co.

Figure 11-3 Open Office (Bull Pens)

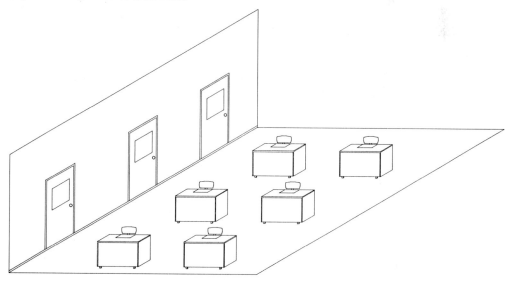

7. *Files* and *literature* are accessible to all requiring less files and copies of magazines and journals.

8. *Cleaning, vacuuming, and sweeping* work is reduced.

The disadvantages of the open office concept are:

1. The *lack of privacy* is probably the biggest problem with open offices. Fellow employees can, very innocently, interrupt the thinking or concentration on a difficult task which then requires starting over. If people are too close and accessible to each other, non-business discussions can eat up great amounts of time. This process is called *coffee clotching*. It can decrease productivity and quality and must be discouraged.

2. *Noise* is another problem with open offices. Equipment which produces most of the noise can be isolated for noise control. But, open offices are more noisy than private offices.

3. Open office space doesn't have the *status* that a private office carries. The recruitment of a good potential employee may be missed because of the office space quality.

4. *Confidentiality* of some work may require private space.

The choice of open office space or private office space depends upon balancing the advantages and disadvantages for each position. Each company could have both open and private offices, but who gets a private office is in important decision that cannot be made without high-level planning.

Conventional Offices

Conventional offices (see Figure 11-4), also known as *fixed walled offices,* are the opposite of open offices. A conventional office has independent furniture, four walls, and a door. More than one person can be assigned to an office, and at what point it becomes an open office is a little fuzzy to most layout technicians, but if more than one function is performed in this space, it's an open office. A *function* may be accounting, purchasing, personnel, engineering, data processing, sales, or production. Conventional office layouts are older than open offices, but both can be improved upon. A combination of open office concept and conventional office advantages would allow for the best of both techniques. We will call this the *modern office concept.*

The Modern Office

The modern office design concept (sees Figures 11-5 and 11-6) tailors individual work areas to satisfy the needs of the organization. The modern office will provide private office space where needed without negatively affecting the cost of utilities, maintenance, and accessibility. Figure 11-5 is a picture of a modern office and Figure 11-6 is a typical layout. Notice the equipment.

Figure 11-4 Conventional Offices

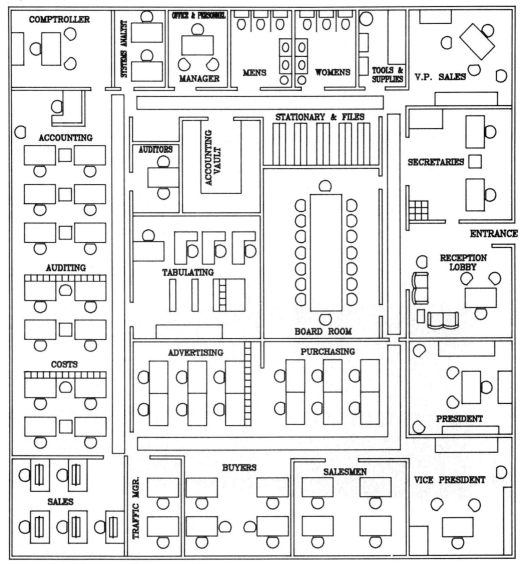

Courtesy of Global Equipment Co.

1. Panels do not go either to the ceiling or the floor. Air can circulate. Panels are padded with soft material to hold down noise.
2. Cabinets are built into the panels to better use the space over desks and tables.
3. Tables are built into the panels to save space and costs.
4. Drawers under the tables allow for storage of supplies just as a desk would.
5. Utilities (electrical, computer, and phone) lines can be carried in the panel. This will give the office a cleaner look and it will also improve safety.

Modern offices can be arranged and rearranged to meet the changing needs of the organization. Modern organizations are developing teams to solve problems. These problems and the makeup of teams keep changing, but the office needs must keep up. The modern office is very flexible. When the company moves, the walls move too.

Figure 11-5a Modern Office

Courtesy of American Seating.

Figure 11-5b Modern Office

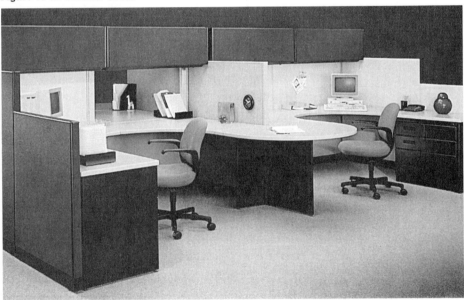

Courtesy of American Seating.

Figure 11-5c Modern Office

Courtesy of American Seating.

Modern offices have been described using various terms like *clustered of-fices, landscaped offices,* and *free-standing offices.* What ever the term, the purpose of modern office design techniques is to eliminate the disadvantages of open offices and traditional offices, and to promote cost effectiveness in the long run. Figure 11-7 shows a comparison between conventional and modern office space.

Modern offices should be pleasing to look at, convenient for its users, comfortable, and efficient. The justifications vary from employee relations to customer opinion development to cost consciousness.

SPECIAL REQUIREMENTS AND CONSIDERATIONS OF SOME OFFICES

Keep these points in mind when designing your offices:

1. *Privacy* may be required by some office employees. Personnel problems should be discussed in private. Many financial matters are confidential. Corporate planning can consider many alternatives that will never come to pass, so to prevent harmful rumors, privacy is needed.

Figure 11-6 Modern Office Layout

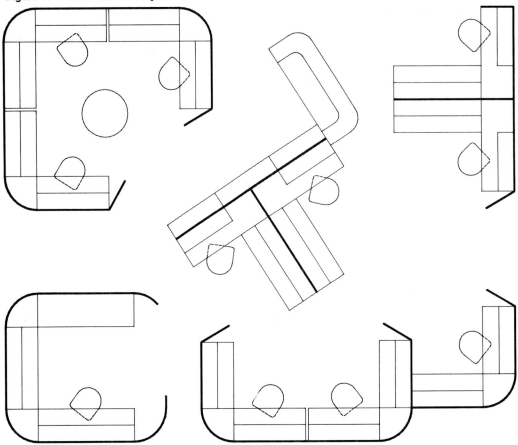

2. *Point of use storage* (see Figure 11-8) is a layout principle that requires the storage of supplies close to the point of use. Office supplies vary from department to department. Engineering supplies are not the same as accounting supplies; personnel forms are not at all like purchasing forms. Therefore, every office department needs a supply room or controlled area. In small offices, a desk drawer may be used, but in large facilities, large controlled areas are needed to handle these valuable commodities.

3. Offices in manufacturing plants often have a *second floor*. Typically, an office is built within the manufacturing plant. The ceilings of a manufacturing plant are often 20 feet or higher, so using only one floor is a waste of building cube. Building a second floor is good cube utilization. The functions of the department placed on the second floor should not require outside visitors or much traveling during the day. Personnel, purchasing, and sales have a lot of visitors, so they would be on the ground floor. Engineering, accounting, market-

Figure 11-7 Conventional Space vs. Modern Space—10 Keypunch Operators' Workstations

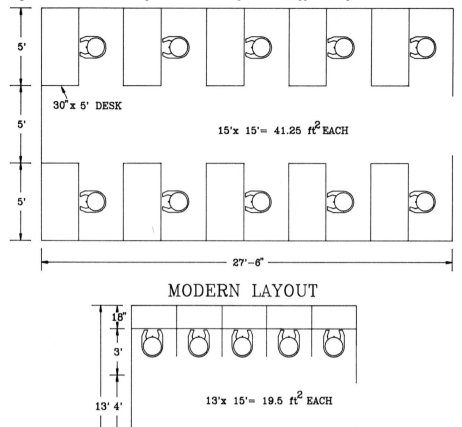

30" x 5' DESK

5'

5'

5'

15'x 15'= 41.25 ft^2 EACH

27'–6"

MODERN LAYOUT

18"

3'

13' 4'

13'x 15'= 19.5 ft^2 EACH

15'

NOTE:
    At a rental rate of $25.00 per square foot per year the conventional space would cost $10,312.50, while the modern layout would cost $5,625.00 per year, a savings of $4,687.50 per year.

ing research, data processing or order entry (phone) do not have as many visitors, so they could be on the second floor.

4. *Centralized or decentralized?* Where do we place offices? Where they are needed may be the best answer. The question usually is, do we have one big

Figure 11-8 Point of Use Storage

Courtesy of White Aisle Saver

office in the front of the building (centralized) or several smaller offices throughout the plant. The advantages to a centralized office are:

a. single office area construction including a common air conditioning system and block wall;

b. the convenience of having all office people in one area;

c. convenience to outside visitors without disturbing production; and

d. common files and equipment.

The disadvantages of a centralized office is that it is not convenient for other operation departments like receiving, shipping, maintenance, stores, warehousing and production, who all have important relationships with the office.

5. Office *flexibility* is an important consideration from the very beginning. When building an office, immediate consideration should be given to expansion. Footings and supports for a second story will be much cheaper if they are put in in the beginning. If we have to break up floors and walls in order to build a second floor, we'll never expand upward. Walls must be flexible also. Most offices will grow. Partitions and panels are better (more flexible) than walls for this reason. Utility flexibility is also important. Several methods are used to create utility flexibility.

    a. *Q floors* are like corrugated metal placed on the ground before the concrete for office floor is poured. The Q floor allows for the running of electricity, computer cable, telephone lines, and the like every fourth foot for the full length of the office (see Figure 11-9). If a desk is moved, the old plug ins can be plugged up and new plug ins added.

    b. *Drop ceilings* and hollow panel supports keep the utility cords and cables concealed overhead in the ceiling. They can be dropped anywhere.

6. *Conference rooms* can be used to provide privacy when needed in open office areas. Privacy may be needed by supervisors holding disciplinary sessions or by salespeople with customers. The important thing is that privacy is available in open office layouts. Board rooms are special conference rooms where the board of directors of a public corporation meet and, therefore, must be laid out for privacy and noise reduction.

7. *Libraries* are special needs areas where reference books and magazines are kept. This is a cost reduction idea. Instead of purchasing books for individuals to put in their book cases, books are purchased by the library and kept in a central, convenient area. A *Wall Street Journal* subscription costs over $100 per year. If ten managers can share one copy, $900 per year can be saved. The paper is then available to everyone else as well. Professional journals, handbooks, and many other publications all add to the value of a reference library.

8. A *reception area* is the visitor's center. The front door of your company is where visitors enter. A receptionist will greet the guest and ask how they may be helped. While the receptionist searches out assistance, our guest needs a place to wait. This should be a comfortable, attractive area so as to create a

Figure 11-9 Q Flooring

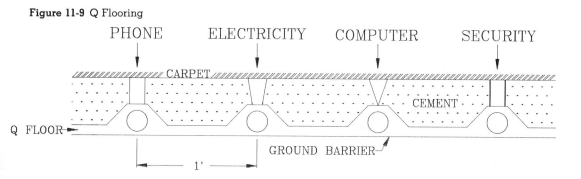

favorable opinion about our company. The best way to make a good opinion is not to keep the guest waiting. But unannounced visitors may need to wait, so this area should be equipped with chairs, desk, telephone, magazines, and company information. Product display may help our visitors visualize how they could help us.

9. *Telephone systems* are becoming automatic, but some personal attention is always needed. If the volume of calls is not too great, the receptionist can handle the telephone as well as the reception area. Telephone equipment requires space. The central board and exchange may be big enough to have their own room, but these areas can be remote and out of the way.

10. *Copy machines* can be major pieces of equipment. This equipment needs special material, operation instructions, and a clean environment. Like every other piece of equipment, a workstation layout is needed. Smaller copiers may be a part of a small office, but big equipment is a department of its own. Storage areas, work in progress areas, and finished work storage areas are required.

11. Incoming and outgoing mail can be big business. A company's mail comes into the *mailroom* and is sorted. Mail is then either delivered or is picked up by employees. Outgoing mail will require postage, weighing, and sometimes folding inserts, stuffing envelopes, and sealing closed. This is called *mass mailings*. Special equipment is available to do this automatically, so the mailroom layout will need equipment layouts as well.

12. Companies create and receive many kinds of documents. Legal requirements force companies to keep many of these documents for years. This creates a need for *file storage areas*. Also, blueprints, processing information, purchase orders, and the like are needed by many people. Central files reduce the need for many copies. Computers and microfilm are reducing the space requirements and the configuration of file rooms, but file rooms are still needed.

13. A *word processing pool* is a group of secretaries in a central area who receive work from many sources. This is an alternative to private secretaries and is, in general, a more efficient use of people.

14. *Aisles* are big users of space. In open offices, the smallest aisles are 3 to 5' and the larger aisles are 6 to 8'. The traffic during peak periods will determine aisle sizes.

15. More equipment and systems are being controlled by *computer* every day. Main frames and central processing units are kept in special temperature/humidity controlled rooms. Computer security is also important.

16. Other areas and considerations to keep in mind are:
    a. lighting;
    b. vaults;
    c. standardization; and
    d. expansion.

The techniques used for creating an office layout are:

1. The organizational chart
2. Procedures diagram (systems and procedures analysis)
3. Communications force diagram
4. Activity relationship diagram
5. Worksheet
6. Dimensionless block diagram
7. Size development
8. Detailed master plan

Analyzing organizational needs, paperwork flow, who works with whom, and the relationship among departments leads to a master plan. Each technique will be described in detail in this section. Follow the procedures given and an efficient and effective layout will result.

The Organizational Chart

The *organizational chart* (see Figure 11-10) gives the layout technologist an idea of the office size. The organizational chart tells us how many people work in each area and level of the company. Each department must be considered and the space determined. The total number of people is the best indication of office size required. A rough estimate of office space needs can be calculated by multiplying the number of people requiring office space by 200 square feet each. So, for example, the overall size of an office for 100 people would be 20,000 square feet. This would include everything. This is a good initial planning tool, but should only be used for determining the total office space, not the departmental space.

Who works in the office, who reports to whom, how many people are in a department, what functions are performed, and similar questions are all covered in the organizational chart. Determining the number of people at each level of the company is another way of calculating space.

Employees	Square Feet
General Managers and Senior Executives	200–300
Managers	150–250
Supervisors	100–200
Accountants	75–150
Engineers	100–150
Clerks	75–100

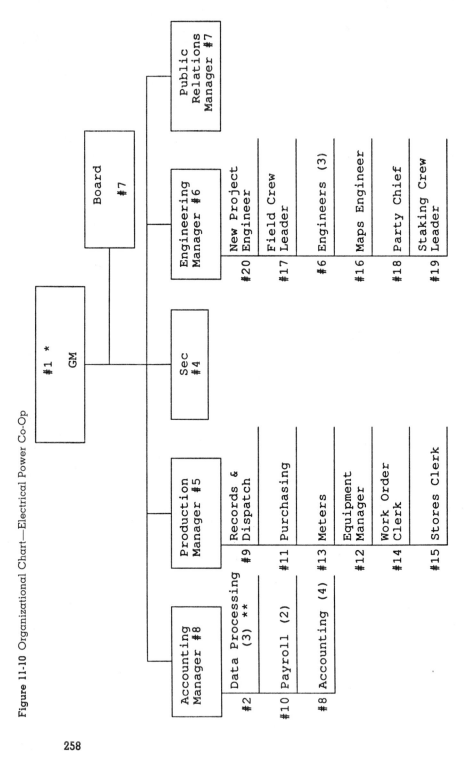

Figure 11-10 Organizational Chart—Electrical Power Co-Op

* Position Numbers (#1)
** Number of People in This Office (3)

258

Procedures Diagram

Procedures diagramming is very much like flow process charting, but instead of following the flow of a product, we follow the flow of every copy of a form. Sometimes, we follow many forms (like the example in Figure 11-12) because one form causes the creation of another form and so on and so on. To analyze paperwork flow, the *procedures diagram* technique was developed. Standard symbols (process chart symbols) have been developed to help explain the standard steps like those shown in Figure 11-11.

Figure 11-12 illustrates the purchase order procedure. The people or departments are listed down the side. Starting with the requestor asking for something (submitting a purchaing request) and receiving approvals, the purchase order is created and the copies sent to four other areas (one copy is filed with the requisition in an open file in purchasing). One copy goes to the requestor, one to accounting, one to receiving, and one copy to the vendor supplying the item. Once the

Figure 11-11 Process Diagram Symbols

Operation	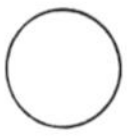 = ◯	Perform some function like match, review, fill orders, input data, etc.
Form	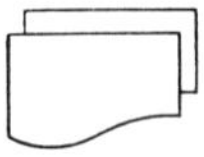 = ▱	Generate a form. A special operation. If more than one copy is used another page is shown behind the first page for each copy.
File	 = ▽	File documents. A "T" could be placed inside the triangle to indicate temporary file or follow up file and a "P" could be used for permanent file or completed file.
Transportation	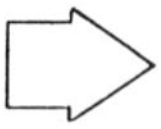 = ▷	Physically moving something like material (not paperwork).
Decision	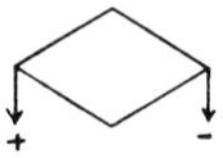 = ◇	Yes/No, Go/No Go at any point where the direction of flow might change.
Approval	= ☐	Used when a management approval is required.
Paperwork Flow	 = ⟶	Shows flow of information.
Telephone	 = ⌐	Shows flow of information by telephone or computer.
Processing	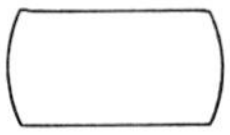 = ▭	Used for computer processing.

Figure 11-12 Purchase Order Payment System

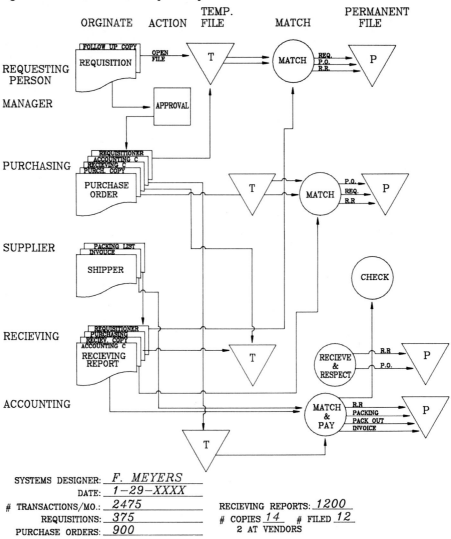

SYSTEMS DESIGNER: *F. MEYERS*
DATE: *1–29–XXXX*
TRANSACTIONS/MO.: *2475*
REQUISITIONS: *375*
PURCHASE ORDERS: *900*

RECIEVING REPORTS: *1200*
COPIES *14*   # FILED *12*
2 AT VENDORS

order is shipped and received, a copy of the vendor's packing list is matched up with the purchase order and a receiving report is made. Five copies of purchase order, two copies of requisitions, four copies of receiving reports, a packing list, and an invoice all have to end up in the files.

Figure 11-12 shows the movement of purchase order forms around the office. This movement has an affect on the office layout. When all the forms are analyzed, relationships between departments will become clearer and relationship codes can be developed. A from-to chart could be developed resulting in a most efficient layout. The from-to chart has not been included in this section but it could be the best tool for optimizing paperwork flow.

Communications Force Diagram

The communications force diagram is another way of determining office relationships (see Figures 11-13 and 11-14). The procedures diagram method requires analyzing all paperwork flow and diagramming all the procedures. This can be a job so big that years of analysis may be needed. The results of procedures diagramming will be extremely valuable, but for our office layout needs, the communications force diagram is much faster.

The communications force diagram requires the technologist to talk with each person involved in the office and find out who they work with most. Each person we talk with will be the center of the diagram, and each person they work with is placed on the periphery (see Figure 11-13). The number of lines connecting the subject person to the periphery people will indicate the importance of the relationship as follows:

1. If there are *4 lines,* it is absolutely necessary that these two people be close together. This code should be reserved for people who communicate several times an hour. This will be coded an "A" relationship.
2. If there are *3 lines,* it is especially important that these two people be close to each other. This code should be reserved for people who need to communicate with each other at least once an hour. This will be coded as an "E" relationship.

Figure 11-13 Communication Force Diagram—General Manager

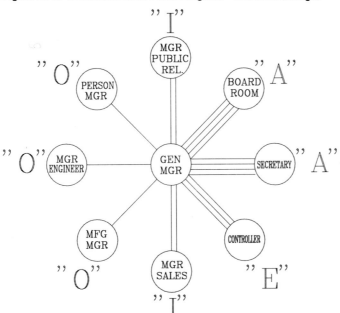

Figure 11-14 Communications Force Diagram Electrical Power Coop

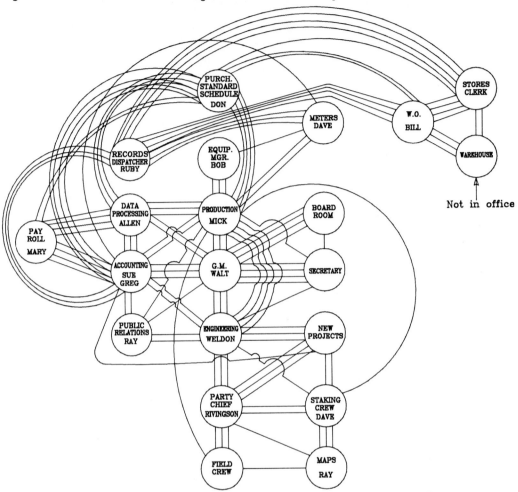

3. If there are *2 lines,* this is an important relationship and these two people should be close. This relationship is reserved for people working with each other several times a day. This will be coded an "I" relationship.

4. If there is *1 line,* this is an ordinary relationship and is reserved for people who interface on a daily basis. This will be coded an "O" relationship.

Figure 11-13 is a communications force diagram for one person. With an office of 22 people, 22 communications force diagrams will be needed. These 22 diagrams must be summarized into one large diagram with 22 circles with all the lines between every circle. Figure 11-14 is an example of this. Notice the *long* lines. These are departments we need to put closer together. Those people or departments that have the most contact outside the office are on the perimeter of the diagram; people with a lot of contact within the office are located in the middle of the diagram. The relationships established here will be carried forward to the activity relationship diagram.

Activity Relationship Diagram

The activity relationship diagram was discussed in Chapter 5. In brief, it shows the relationship of every department or person with every other department or person. A simple code (A E I O U X) is used to tell the importance of the relationship (see Figure 11-15). In the office layout, we use the communication force diagram to establish these importance codes or we can talk to each department or person included in the study and have them record the codes. Figure 11-15 was developed from the communications force diagram (see Figure 11-14).

Figure 11-15 Activity Relationship Diagram

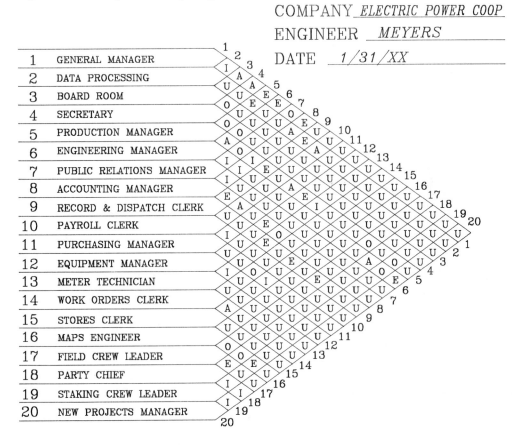

COMPANY _ELECTRIC POWER COOP_

ENGINEER _MEYERS_

DATE _1/31/XX_

1	GENERAL MANAGER
2	DATA PROCESSING
3	BOARD ROOM
4	SECRETARY
5	PRODUCTION MANAGER
6	ENGINEERING MANAGER
7	PUBLIC RELATIONS MANAGER
8	ACCOUNTING MANAGER
9	RECORD & DISPATCH CLERK
10	PAYROLL CLERK
11	PURCHASING MANAGER
12	EQUIPMENT MANAGER
13	METER TECHNICIAN
14	WORK ORDERS CLERK
15	STORES CLERK
16	MAPS ENGINEER
17	FIELD CREW LEADER
18	PARTY CHIEF
19	STAKING CREW LEADER
20	NEW PROJECTS MANAGER

Activity Worksheet (Figure 11-16)

The *activity worksheet* was also talked about in Chapter 5. Figure 11-16 was made from Figure 11-15 in order to create 20 individual blocks. This worksheet moves us from the activity relationship diagram to the dimensionless block diagram.

Dimensionless Block Diagram

To create a *dimensionless block diagram,* cut out 20 square paper blocks of about 2″ × 2″. On the worksheet, starting at line one, place the number of the line and the name of the department in the middle of the block (see Figure 11-17). Then, starting at the top left-hand corner, place the ''A'' relationships this department has with other departments. For example:

Figure 11-16 Worksheet for Activity Relationship Diagram—Electric Coop

	ACTIVITY		DEGREE OF CLOSENESS				
		A	E	I	O	U	X
1.	GENERAL MANAGER	3,4	5,6,8	2	7		
2.	DATA PROCESSING	8,10	5,9	1	—		
3.	BOARD ROOM	1	—	—	4		
4.	SECRETARY	1	—	—	3,5,6		
5.	PRODUCTION MANAGER	6,11	1,2,9,12	8,13	4,7,17,19		
6.	ENGINEERING MANAGER	5,18	1,20	7,8	4.19		
7.	PUBLIC RELATIONS MANAGER	—	—	6,8	1,5		
8.	ACCOUNTING MANAGER	2,10	1,9,12	5,6,7	13		
9.	RECORDS & DISPATCH	—	2,5,8,13,14,17	—	—		
10.	PAYROLL CLERK	2,8	—	11	—		
11.	PURCHASING CLERK	5	—	10,14	15		
12.	EQUIPMENT MANAGER	—	5,8	13	—		
13.	METER TECHNICIAN	—	9	5,12	8		
14.	WORK ORDER CLERK	15	—	11	—		
15.	STORES CLERK	14	9	—	11		
16.	MAPS ENGINEER	—	19	—	17,18		
17.	FIELD CREW LEADER	—	9,18	—	5,16		
18.	PARTY CHIEF	6	17	19,20	16		
19.	STAKING CREW LEADER	—	16	18,20	5,6		
20.	NEW PROJECTS MANAGER	—	6	18,19	—		

Now place the "E" relationships on the top right corner, the "I" relationships in the bottom left, and the "O" relationships on the bottom right. Finish all 20 blocks.

When all 20 blocks are completed, find the one block with the most important (A's and E's) relationships and place it in the middle of your desk. For example:

	6	
6,11		1,2,9,12
11	5 Production Manager	1
8,13		4,7,17,19
12	8	4

Now place the most important offices around this center office until the "A" relationships are satisfied. When you place offices 6 and 11, they will have "A" relationships to satisfy. Keep working with the "A" relationships until all offices with an "A" relationship have a full side contact with each other. Now you can start working with the "E" relationships, the "I" relationships, and finally the "O" relationships. We try to accommodate the "O" and "I" relationships, but they are often prevented from being close by so many other important relationships. You may try many layouts. Be sure you keep track (by making a small plot plan) of your layouts before going on to a change.

Figure 11-17 Dimensionless Block Diagram—Electrical Power Coop Office

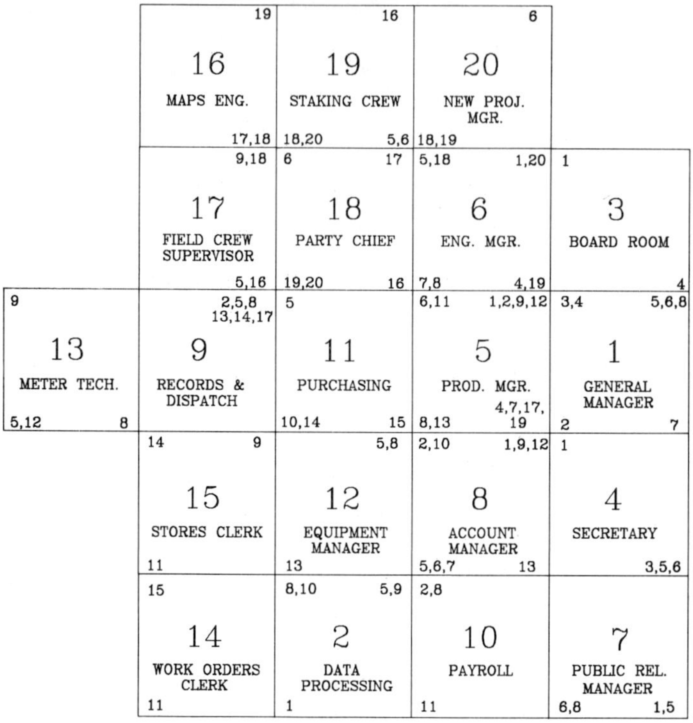

```
| 19            | 16            | 6                |
|   16          |   19          |   20             |
| MAPS ENG.     | STAKING CREW  | NEW PROJ.        |
|               |               |    MGR.          |
|        17,18  | 18,20   5,6   | 18,19            |
|----------------------------------------------------------------|
| 9,18  |6              17| 5,18       1,20 | 1             |
|   17          |   18          |   6              |   3          |
| FIELD CREW    | PARTY CHIEF   | ENG. MGR.        | BOARD ROOM   |
| SUPERVISOR    |               |                  |              |
|         5,16  | 19,20      16 | 7,8        4,19  |           4  |
|-----------------------------------------------------------------------|
| 9     | 2,5,8       |5            | 6,11    1,2,9,12 | 3,4     5,6,8|
|       | 13,14,17    |             |                  |              |
|  13           |   9           |   11          |   5          |   1         |
| METER TECH.   | RECORDS &     | PURCHASING    | PROD. MGR.   | GENERAL     |
|               | DISPATCH      |               |     4,7,17,  | MANAGER     |
| 5,12    8     | 10,14     15  | 8,13       19 | 2         7  |             |
|-----------------------------------------------------------------------|
| 14         9  |           5,8 | 2,10    1,9,12| 1            |
|   15          |   12          |   8           |   4          |
| STORES CLERK  | EQUIPMENT     | ACCOUNT       | SECRETARY    |
|               | MANAGER       | MANAGER       |              |
| 11            | 13            | 5,6,7      13 |        3,5,6 |
|-----------------------------------------------------------------------|
| 15            | 8,10     5,9  | 2,8          | 
|   14          |   2           |   10         |   7          |
| WORK ORDERS   | DATA          | PAYROLL      | PUBLIC REL.  |
| CLERK         | PROCESSING    |              | MANAGER      |
| 11            | 1             | 11           | 6,8     1,5  |
```

There are hundreds of possibilities, and the one best answer is the one that satisfies the most relationships.

Once we have the final dimensionless block diagram, we can identify where outside walls will be, where the shops, warehouses or stores departments (non-office) so an orientation can be developed. This will be our plan for where each office goes. We still need size.

Size Development

To calculate the space requirement for the office, we can use the following techniques:

1. *The 200 square feet per person technique.* The 200 square feet per person technique is used for establishing the overall office space. If we look at Figure 11-10 (the organization chart for the Electrical Power Coop) we count 36 people requiring office space, so 36 × 200 = 7,200 square feet. So, our office needs to be 7,200 square feet.

2. *The level of the organization technique.* Looking at Figure 11-10, we find the following information:

# People	Positions	Sq. Ft. Each	Total Sq. Ft.
1	Senior Exec.	250	250
4	Managers	200	800
4	Clerical	100	400
9	Accounting	100	900
5	Engineers	125	625
6	Supervisors	150	750
	100% extra space allowance		3,725
		Total	7,450 ft.²

So, we need a total of 7,450 square feet.

3. *The workstation technique.* The workstation layout approach is the most detailed and will include restrooms, lockers, cafeterias, reception areas, board rooms, conference rooms, and anything else that takes up space. This technique adds 25 percent extra space for expansion. We do not require a full layout before construction begins. Square footage of an office is simply the length multiplied by the width of the office measured in feet. A 20′ × 20′ office would be 400 square feet.

Detailed Master Layout

No matter which technique we use, we need to known the length and width of the office for our layout. These measurements may be enough for construction to begin, but more detail will be required. The department layouts are the next level of detail, which will include the internal walls or department boundaries. The final level of detail is where desks, chairs, and all the other equipment goes. This detailed plan will be needed before assignments of space can be achieved. A good guide for space planning and layout is Federal Stock No. 7610-145-0168, Page 10.6.1. Write to the Superintendent of Documents, Washington, DC.

In creating your detailed master layout, keep the following rules in mind:

- Desks should face the same general direction.
- In open areas, desks should be placed in rows of two.
- For desks in one row, there should be 6′ (1.83m) from the front of one desk to the front of the desk behind it.
- For desks in rows of two or more and where ingress and egress is confined to one side, 7′ (2.13m) should be allowed from the front of one desk to the front of the desk behind it.
- If employees are back to back, allow a minimum of 4′ (1.22m) between chairs.
- Inside aisles within desk areas should be 3 to 5′ (0.91 to 1.52m) wide.
- Intermediate aisles should be 4′ (1.22m) wide.
- Main aisles should be at least 5′ (1.52m) wide.
- Natural lighting should come over the left shoulder or the back of an employee.

- From 50 to 75 square feet (4.65 to 6.97m²) are required for a work space consisting of a desk, shelf space, a chair, with a 2′ (0.61m) space allowance on the length and width.
- Desks should not face high-activity aisles and areas.
- Desks of employees doing confidential work should not be near entrances.
- Desks of employees having much visitor contact should be near entrances with extra space provided.
- Desk of the receptionist should be near the visitor's entrance.
- Supervisors should be positioned adjacent to the secretaries.
- Supervisors in open areas should be separated from their group by 3.3′ (1m).
- The flow of work should take the shortest distance.
- People who have frequent face-to-face conferences should be located near each other.
- Employees should be adjacent to those files and references they use frequently.
- Employees should be placed near their supervisors.
- Five-drawer file cabinets should be considered in lieu of four-drawer cabinets.
- Open-shelf filing or lateral file cabinets should be considered in lieu of standard file cabinets.
- Four- or five-drawer file cabinets should be considered as a substitute for two two-drawer cabinets.
- The reception area should create a good impression on visitors and an allowance of 10 square feet (0.93m²) should be used per visitor if more than one arrives at a given time.
- The layout should have a minimum of offsets and angles.
- Large open areas should be used instead of several small areas.
- Open areas for more than 50 persons should be subdivided by use of file cabinets, shelving, railings, or low "bank-type" partitions.
- Office space should not be used for bulk storage or for storage of inactive files.
- Conference space should be provided in rooms rather than in private offices.
- Conference and training rooms should be pooled.
- The size of a private office will often be determined by existing partitions.
- Private offices should have a minimum of 100 square feet (9.3m²) to a maximum of 300 square feet (27.9 m²).
- A 300 square feet (27.9m²) private office should be used only if the occupant will confer with groups of eight or more people at least once per day.
- Related groups and departments should be placed near each other.
- Minor activities should be grouped around major ones.
- Work should come to the employees.
- Water fountains should be in plain view.

- Layouts should be arranged to control traffic flow.
- Heavy equipment generally should be placed against walls or columns.
- Noise producing workstations should be grouped together.
- Access to exits, corridors, stairways, and fire extinguisher should not be obstructed.
- All governmental safety codes should be followed.
- In planning the office, consider the floor load, columns, window spacing, heating, air conditioning and ventilation ducts, electrical outlets, and lighting and sound.
- The scale of the layout should be either $\frac{1}{4}'' = 1'$ (1 cm = 50 cm) or $\frac{1}{8}'' = 1'$ (1 cm = 100 cm).
- Plastic reproducible grid sheets and plastic self-adhesive templates should be considered.

QUESTIONS

1. What are the goals of office layout?
2. What are the four types of office space?
3. What are the advantages of the open office layout concept?
4. What are the disadvantages of the open office layout concept?
5. List 19 special office requirements and considerations.
6. What are the techniques of office layout?
7. How does the organization chart help in office layout?
8. How much space is required in the office (rough estimate)?
9. What are the standard symbols of the procedures diagram?
10. What is a communications force diagram?
11. What symbols are used in the communications force diagram?
12. What is the basic source of information to create an activity relationship chart?

Chapter 12

Area Allocation

Area allocation is a process of simply dividing up the building's space or to allocate space among the departments. To allocate space, of course, you need to know how much space is required. This book has been developing space requirements for a tool box plant since Chapter 3. Let's continue with that example in order to illustrate area allocation.

SPACE REQUIREMENTS PLANNING: STEP ONE

A total plant size and shape is needed very early in the project in order to design the building. Each department's space needs are analyzed and listed on a *total space requirements worksheet*. The manufacturing space (Chapters 3 and 6), production services space (Chapter 7), employee services space (Chapter 8), office space (Chapter 11), and outside area space (Chapters 7 and 8) are all determined separately and then listed on the worksheet. Figure 12-1 is a recap of the space requirements for our tool box plant. The number in parenthesis after the length and width dimensions on the total space requirements worksheet are figure numbers or paragraph numbers where these space requirements come from. In a summary form, it is very important that each figure be backed up with design data so no one will think you just pulled this space requirement out of thin air. That would be a big mistake. All space requirements are based on something, so be sure to reference where every size came from.

The space requirement for the fabrication area is a total of all machines and workstations. The area for one machine is the maximum length times the maxi-

Figure 12-1 Total Space Requirements Worksheet
Tool Box Plant

	Stations ×	W ×	L	(Figure #) =	Square Feet
I. MANUFACTURING					
A. Fabrication					
Strip Shear	2 ×	9.5 ×	12	(6-3)	228
Chop Shear	4 ×	7 ×	13	(6-3)	364
Punch Press	3 ×	8 ×	11	(6-3)	264
Press Break	6 ×	8 ×	11	(6-3)	528
Roll Former	1 ×	6 ×	18	(6-3)	108
Fabrication TOTAL:					1,492
B. Spot Weld	1 ×	26 ×	30	(3-7)	780
C. Paint	1 ×	28 ×	100	(6-4)	2,800
D. Assembly & P.O.	1 ×	16 ×	38	(3-8)	608
Subtotal					5,680
50% Allowance (mostly aisles)					2,840
Manufacturing Total:					8,520
II. PRODUCTION SERVICES					
Receiving—Steel		13 ×	25	(7-2)	325
Receiving—Cartons		17 ×	19	(7-3)	323
Stores		18 ×	25	(7-28b)	450
Warehouse		64 ×	68	(7-16)	4,352
Shipping		20 ×	20	(7-6)	400
Maint. & Tool Room	(2 people @ 400 ft. ea.)				800
Utilities	(Estimate only)*				100
Production Services Area Total: (aisles are included in each layout in this area.)					6,750
III. EMPLOYEE SERVICES					
Employee Entrance		10 ×	20	(8-3)	200
Locker Room	(3.5 ft.²/employee × 50 employees)			(8-5)	175
Toilets		10 ×	20	(8-7)	200
Cafeteria	(10 ft.²/employee × 50 employees)			(8-9)	500
Drinking Fountain	(6 fountains × 15 ft.² each)			(8-G)	90
Medical Services	(First Aid Room Only 10′ × 10′)			(8-1)	100
Services Area Required Total:					1,265
IV. OFFICE AREA (11 people from organizational chart)					
(11 people × 200 ft.² each (11-D-1)					2,200
TOTAL BUILDING SPACE					18,735

V. OUTSIDE AREAS
Receiving, Parking and Maneuvering Area
Shipping Parking and Maneuvering Area
Employee Parking (50 employees)
   1.5 employees per parking space
   250 ft.²/parking place (8-2)

$$\frac{50 \text{ employees}}{1.5 \text{ employees/spaces}} = 34 \text{ spaces}$$

34 spaces × 250 ft²/space = 8,500 ft²

---

* I didn't size it properly, but the error will be minimized by an educated guess. This is a very minor use of area.

mum width. This makes a rectangle out of each machine and space may be saved by fitting irregular-shaped workstations or machines more creatively. Any space saved in this way can be used in expansion plans for the future. Also, it is nice to have a little extra space because the most common error of plant layout is omission (we forgot something). The size and shape of a department may change to fit into our final building shape. The size should be very close because we minimized space needs while designing that department, but the shape almost always changes a little to fit with other departments in our newly designed plant shape.

Before converting the space requirements from Figure 12-1 into plant space, the cube utilization must be reviewed. Most of our layout design has concentrated on floor space, but not everything needs to be placed on the floor. Other levels within the plant may be suitable. Consider the following areas.

Under the floor. Basements are the biggest user of under the floor space. Almost anything can be placed in a basement area. Walkways can also be placed underground, especially between buildings. The disadvantages to basement areas are additional construction costs, stairs (safety), elevators (flow restrictions), and maintenance costs. But utilities (electrical, compressed air and water) can be placed under the floor in small trenches keeping the overhead areas clear for material handling equipment. This is a cost saver.

Overhead. Overhead areas can be used in many different ways.

Clear space. Clear space is that space from 8' above the floor to the ceiling (also called the *truss*). If a building has 22' high ceilings and we use shelves that stack material only 6' high, we have only used 27 percent of the available height. A mezzanine could more than double this utilization. A stepladder and 8' shelves would increase our utilization even more. Racks commonly use the full height of the building. In the paint department, we can stack two dryers on top of each other and move material by overhead trolley conveyor. Overhead conveyor movement of material is a good use of the building cube in manufacturing. We have made good use of building cube in warehousing, stores, paint and manufacturing, but what about locker rooms, restrooms, cafeterias and offices?

If we could place locker rooms over restrooms, we could save floor space. If we could build a two-story office, we could cut the floor space in half. Building second floors is more expensive per square foot than building on the ground level, so a balance of land cost and construction costs must be made with operations efficiency and safety. One floor is preferred in most situations.

Truss level. A *truss* is a rafter. The size of the space in the trusses depends upon the width of a bay. The wider the bay (span) the thicker the truss. Trusses vary from 2' to 10'. Depending on the size of the truss, many things can be placed in this area. Offices (supported on the floor) are built in the trusses of aircraft plants. Walkways are built in the trusses of steel mills. Many plants run utilities in the trusses. Heaters, blowers, sprinklers, ovens, and the like can be placed in the trusses.

Roof. The roof, although not inside the plant, could be used for recreation use, the central air conditioning system, a silo for material storage, water towers, cooling towers, quality control testing, parking, and the like. Anything we can get off the floor will reduce our building size, always review for cube utilization before determining building size.

BUILDING SIZE DETERMINATION

Our building needs to be 18,735 square feet. A standard building is cheaper than custom-designed buildings. No one would build a 18,735 square foot building because it would be too expensive. Standard buildings come in many size increments like 100' × 100', 50' × 50', 40' × 40', and even 25' × 50'. This refers to column spacing, so a 25' × 50' building would come in multiples of 25' in width and 50' increments in length. A rectangular building results. A two-to-one length to width ratio is a very desirable building shape because of material flow and convenient accessibility. Nearly any ratio of length to width is possible (even squares), but we should start with a two-to-one ratio first.

To establish a two-to-one ratio, divide the total number of square feet needed by two (giving us two equal squares). Then take the square root of one half of that figure. Our plant needs 18,735 square feet. Divided by two equals 9,367.5 square feet. The square root of 9,367.5 is 97'. Round 97' up to 100' (making multiples of 25' and 50'). Now we have the size of our building, 100' × 200'. This is two 100' × 100' areas. A square building would be 137' × 137'. Rounding this up to 150' × 150' or 22,500 square feet. 100' × 200' is 20,000 square feet, a 2,500 foot savings. Remember, 150' across a building can put an employee farther from an emergency exit.

The shape of a building is a unique variable where many answers are correct, but a good starting point is a length to width ratio of two to one.

DIMENSIONLESS BLOCK DIAGRAM

We now have the starting size and shape of our building (100' × ? ?') Our problem now returns to area allocation. How are we going to divide ,   1,000 square foot building? The dimensionless block diagram developed in   5 is the orientation layout plan and is now reproduced here as Figure 12-2. The dimensionless block diagram's relationships must be maintained. A common error is the lack of agreement between the dimensionless block diagram and the final detailed layout.

AREA ALLOCATION PROCEDURE

With the space requirements planning worksheet (see Figure 12-1) and the dimensionless block diagram (see Figure 12-2), the building can now be divided up into departments.

Figure 12-2 Dimensionless Block Diagram

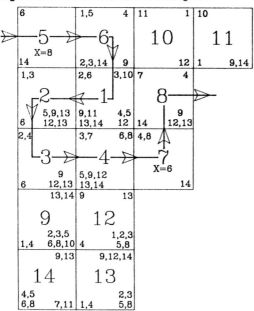

1. The first step in area allocation is to establish a 100′ × 200′ grid using something like $\frac{1}{2}''$ graph paper. A scale of $\frac{1}{2}'' = 20'$ will make each $\frac{1}{2}'' \times \frac{1}{2}''$ square equal to 400 square feet. Figure 12-3a is the first attempt. All that is needed to start are the walls (external only) and the columns (25′ × 50′).

2. The second step of area allocation is to calculate the number of squares (400 square feet) needed by each department.

Department	Square Feet	#400 ft.² Blocks
Fabrication	2,238	6
Spot Weld	1,170	3
Paint	4,200	11
Assembly & P.O.	912	3
Receiving	648	2
Stores	450	1
Warehouse	4,352	11
Shipping	400	1
Maint. & T.R.	800	2
Utilities	100	$\frac{1}{4}$
Emp. Ent.	200	$\frac{1}{2}$
		1
Locker Room	175	
Toilets	200	$\frac{1}{2}$
Cafeteria	500	$1\frac{1}{2}$
Drinking Fountain	—	—
Medical	100	$\frac{1}{4}$
Office	2,200	6
		$49\frac{1}{2}$

We need a total of 50 squares of 400 square feet each or 20,000 square feet. We have assigned all 50 spaces by rounding up.

3. The third step is now to place these blocks into the area allocation layout (Figure 12-3a) using the dimensionless block diagram as the guide. Figure 12-3b shows an allocation of the (50) 400 square feet squares. A few open squares are possible because we built 1300 square feet more than needed, but our example used all 50 blocks because of rounding up. We now know where the departments are going and their shape.

4. Step four of the area allocation procedure is a layout with the internal wall or (better) the area boundaries. Figure 12-3c is the first full plant layout produced

Figure 12-3a 100' × 200' grid (2.5" × 5")
Scale $\frac{1}{2}$" = 20'

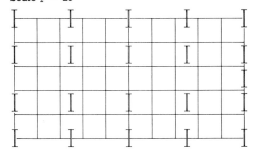

Figure 12-3b Calculate the Number of Squares

R	S	F	F	F	F	UTIL	W	W	W
S.W.	F	F	F	F	F	TR	W	W	W
S.W.	F	F	F	F	F	TR	W	W	SHIP
P	F	F	F	F	ASSY.	P.O.	W	W	W
RR LR	C	O	O	O	O	O			

Figure 12-3c Area Allocation Layout

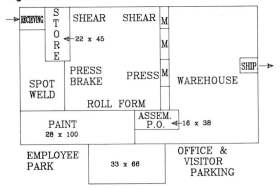

in this book, but there is much detailed work yet to do. The detailed layout with the placement of every piece of equipment will be discussed in the next chapter. Once the area allocation procedure produces a final plan, the architect can now start on building design and construction.

OFFICE AREA ALLOCATION

The office area allocation procedure is the same as the plant area allocation procedure and will allow us a second example. The organization chart (see Figure 11-10) for the electrical power coop office and the dimensionless block diagram (see Figure 11-17) will be our basic source of information for our second problem.

The *organization chart* shows that we need room for 36 people (29 employees and seven board members). The preliminary estimate of 200 square feet of space per person would require 7,200 square feet of office space (200 × 36). The level of the organization technique shows a need for 7,450 square feet. These two figures are very close, which would make any technician very comfortable that about 7,300 square feet of office space will be adequate.

The office size will be:

$$\sqrt{\frac{7200}{2}} = 60 \text{ ft}^2 \text{ or } \sqrt{\frac{7450}{2}} = 61 \text{ ft}^2$$

The building (office) will be 60′ × 120′ (two 60′ × 60′ squares). 60′ × 120′ = 7,200 square feet.

The *dimensionless block diagram,* as discussed in the office layout chapter and as shown in Figure 11-18, is our relationship plan. The closeness relationships were incorporated into the dimensionless block diagram. We must maintain these relationships. These offices and office people referred to in Figures 11-11 and 11-18 are the primary purpose of an office layout. But neither included the personnel services like restrooms, cafeterias, supply stores, or files. Nor do they include office service functions like conference rooms or reception areas.

We considered the seven board of directors space in the number of employee method of analyzing space but not in the level of the organization technique. All these areas plus the aisles are considered in the 100 percent add-on in the level of organization technique.

The area allocation procedure proceeds as follows:

Step 1: Establish a 60′ × 120′ grid using ½″ graph paper. A scale of ½″ = 10′ will make each ½″ × ½″ square equal to 100 square feet. Figure 12-4a is an outline of the office. Allowing 30′ × 40′ column spacing would be a good plan. These columns must be placed on the grid to ensure aisles or equipment is not placed on them.

Step 2: The second step of area allocation is to calculate the number of squares (100 square feet) needed per office or service function. Figure 12-5 is a list of

Figure 12-4a Office Outside Walls—4 Columns (30′ × 40′)

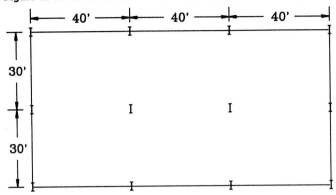

Figure 12-4b Square Allocation

16	16/17	18	3	3	3	6	3	3	3	1	1
17	19	18/19	20	3	3	6	3	3	3		4
	26	26		23	25	21	23	11	5	5	7
13	26	26	23	23	25	21	23	12	8		7
13	14		2		24		23		8	10	8
9	15	2	2	10	10	22	22	22	8	10	8

Figure 12-4c Area Allocation Diagram

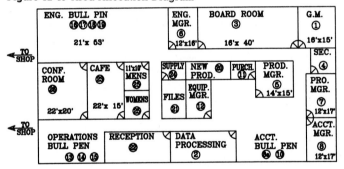

offices as developed from Figure 11-10 (the organization chart). The first number before the name indicates the position number on the organization chart. These position numbers are also used on the dimensionless block diagram. The square feet required for each function came from Chapter 11. Remember, every area space requirement must depend upon something. A number in parenthesis behind the area description indicates the number of people in this

CHAP. 12: AREA ALLOCATION

Figure 12-5 Office Area Space Requirements—Electric Power Coop

Position Number	Area Description	Square Feet Size	No. of 100' Spaces	Approximate Size
1	General Mgr.	250	2.5	15 × 16
2	Data Proc. (3)	300	3	19 × 16
3	Board Rm. (7)	640*	6.5	40 × 16
4	Secretary	100	1	10 × 10
5	Product. Mgr.	200	2	12.5 × 16
6	Engr. Mgr.	200	2	12.5 × 16
6A	Engineers (3)	450	5	28 × 16
7	Pub. Rel. Mgr.	200	2	12.5 × 16
8	Acct. Mgr.	200	2	12.5 × 16
8A	Accounting (4)	400	4	25 × 16
9	Rcrds. & Disp.	100	1	10 × 10
10	Payroll (2)	200	2	12.5 × 16
11	Purchasing	125	1	10 × 12.5
12	Equip. Mgr.	150	2	12.5 × 12.5
13	Meter Tech.	150	1	12.5 × 12.5
14	Work Order Clerk	100	1	10 × 10
15	Stores Clerk	100	1	10 × 10
16	Maps Engineer	125	1	10 × 12.5
17	Field Crew Ldr.	150	1	12.5 × 12.5
18	Party Chief	150	1	12.5 × 12.5
19	Stak. Crew Ldr.	150	2	12.5 × 12.5
20	New Proj. Engr.	150	2	12.5 × 12.5
21	Rstrms. (8-D)(2)	200	2	10 × 10 (2)
22	Reception*	300	3	12 × 25
23	Cafeteria	300	3	15 × 20
24	Stores*	100	1	10 × 10
25	Files*	200	2	10 × 20
26	Cnfrnce. Rm.	400	4	20 × 20
		5,790	61	

* See layout in Figure 12-6.

space if more than one. The total square feet calculated in Figure 12-5 (totals) is only 5,790 and our layout calls for 7,200 square feet. The difference is aisle space. We may be tight (need more space), so aisle space must be used efficiently.

Step 3: The third step (area allocation) starts by placing the dimensionless block diagram (Figure 11-17) and the office area space requirements (Figure 12-5) next to the block diagram (Figure 12-4a). Now, we allocate space by placing the position number(s) in the 100 square feet squares per the dimensionless block diagram. The service areas must be worked in by placing the services conveniently for most of the people. The results of this process will be something like Figure 12-4b. I say ''something like'' because if four of us did this simultaneously, we would have four different but good answers. The main thing is to be true to the dimensionless block diagram.

The final step in the area allocation procedure is to develop a final area allocation diagram. This step requires placing in aisles and specific boundaries.

Aisles should be straight and run the full width and length of an office. Establishing aisles is an important first decision in this last step. This is a small office, so figure on 5' main aisles and 4' cross aisles. Figure 12-4c is the final choice. Four previous layouts were discarded because improvements kept coming. Don't be afraid of trying many different arrangements. The best arrangement satisfies the most relationships as shown on the dimensionless block diagram.

QUESTIONS

1. What is area allocation?
2. What is the total space requirements worksheet?
3. What are the different levels within the plant?
4. How do we convert square footage to building size?
5. What is the area allocation procedure?
6. What is the end result of the area allocation procedure?
7. How can we better use the clear space?
8. Which of these two would you place upstairs and why?
   a. Restrooms or locker rooms
   b. Accounting or purchasing
   c. Old files or current files
9. What is a column? Why is it important?
10. What is column spacing?
11. Using the golden rule of architecture, what would be the length and width of the buildings with the following space requirements:
    a. 825,000 square feet
    b. 250,000 square feet
    c. 87,500 square feet
12. Once the length and width of the building has been determined, how do we know where to place the departments?

Chapter 13

Layout

Layout is a simple term which must communicate the results of many months of data collection and analysis. The layout is only as good as the data backing it up. But it is the visual presentation of your analysis. The term "layout" will be applied to plot plans and master plans. The layout will be your biggest selling tool. As you present your plan to management, you will refer to the layout showing how product flows through the plant. A technique used in Chapter 4 is the flow diagram. The flow diagram cannot be produced until a layout is made. Of course, it could be used on the existing layout as the present method and the basis for cost reduction.

PLOT PLAN

A *plot plan* shows how the building(s), parking lots, and driveways fit on the property (see Figure 13-1) as any other information. Where the main highways, utilities, drains, and the like are as important to the construction project. City and county building codes also affect the plot plan. The driveways (entrances) may require frontage roads and the parking lot may require a set back.

Step 1: Start with a layout of the property showing the lot lines.

Step 2: Place in the main roads that border the property or where the access road will enter the property.

Step 3: Show sources of water, power, gas, and phones.

Figure 13-1 Plot Plan

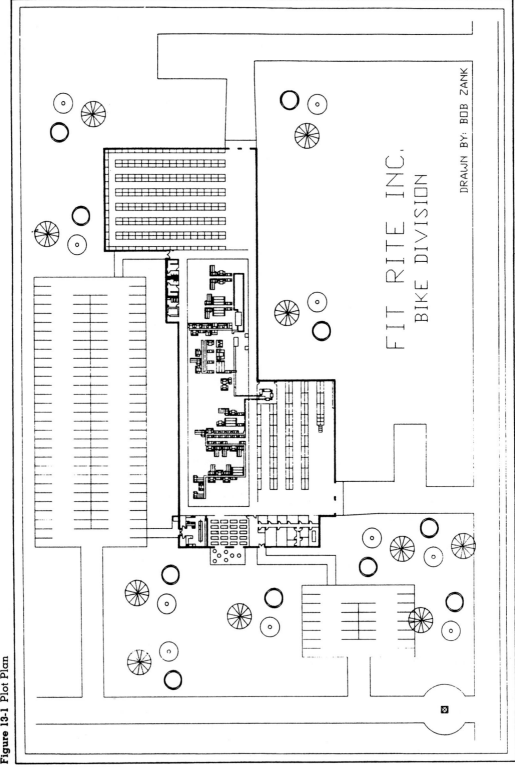

FIT RITE INC.
BIKE DIVISION

DRAWN BY: BOB ZANK

Figure 13-2 Tool Box Plant Plot Plan

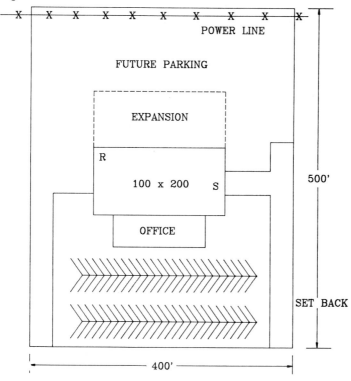

Step 4: Place the building where the front faces the road and the long side faces the road. Expansion plans will go to back.

Step 5: Show receiving and shipping (consider where expansion will go).

Step 6: Connect receiving and shipping to the main road.

Step 7: Show where employee and public entrances will be located.

Step 8: Provide parking for visitor and employees.

Figure 13-2 is the plot plan for our tool box plant.

A plot plan must also show expansion possibilities. It is extremely important to consider expansion even before buying property. Property prices vary with many factors, but one factor that affects the plot plan is frontage cost versus cost of depth of lot. The property on a main road is called *frontage*. The cost of a piece of property will vary proportionally with the size of the front footage. The depth of the lot is not as big a factor. Of course you need adequate size for your plant, but you can purchase additional land behind your lot for a much cheaper price than you can buy on the road. So, we will plan for our expansion behind the building, so be sure not to place any permanent or costly facilities in the way of expansion. Receiving and shipping docks are two facilities that should not be located in the expansion area.

Office expansion may be up (adding a second floor) and the initial construction must allow for a second floor. Parking can be expanded in back, and a new employee entrance may be needed. This will move the locker rooms, restrooms and cafeterias as well, but new facilities will be needed anyway.

When buying property, a general rule of thumb is to buy 10 times more property than the building size. A $100' \times 200'$ building of 20,000 square feet will require 200,000 square feet of land (about 5 acres).

The plot plan communicates a lot of information about how the new plant will fit the lot and what external facilities are required. The architect can now design the driveways, parking lots, and building. The layout technologist can now go back to the internal plant layout problems and create a master plan.

MASTER PLAN

The *master plan* is the finished product of the plant layout and material handling project. When we use the term "layout," we mean master plan most of the time. Every machine, workstation, department, desk, and all other important items are located on the master plan.

Plant Layout Methods

There are four methods which are used for plant layouts:

1. The architectural drawings technique.
2. The templet and tape technique.
3. The three-dimensional models technique.
4. The computer-aided design (CAD) technique.

The architectural drawings technique. The architectural drawing method of developing a master plan is the oldest and most expensive method of producing a plant layout. The need for change in plant layouts is what makes architectural drawings so expensive. If you have just spent a half hour drawing a machine on the layout, and where you drew it is wrong by a foot, or turned 90° from a perfect position, what are the chances you will change that layout. Taking pencil to paper using "T" squares and drawing tools to prepare a layout has been nearly eliminated due to advanced computer software programs. But if this layout project is a once in a lifetime job, then the architectural drawing method may be the right choice.

The templet and tape method. Before the introduction of CAD, the templet and tape method was the preferred technique of technicians who did a lot of plant layouts. The templet and tape technique is a layout made of transparent templets and rolls of various tapes placed on a mylar (plastic) grid base. The mylar has a $\frac{1}{2}''$ grid lightly printed in blue to allow the technician to place walls, aisles, and machines without the use of straight edge or ruler. The wall tape is placed first creating an outline of the building. A $\frac{3}{8}''$ to $\frac{1}{4}''$ tape is used for walls and $\frac{1}{8}''$ tapes are used for aisles. Other tapes are used for air lines, power lines, conveyors (overhead and belt), column posts, and even rolls of operator templets. Figure 13-3 shows a collection of tapes available from the Chartpak Co.

Templets are transparent (sometimes translucent) plastic outlines of very specific pieces of equipment. Figure 13-4 is an example of several machines and office equipment available from Plan Print Corp. Templets can be made by drawing the outline of equipment on white bond paper and then typing on the bond paper the description and size of the equipment. Then make a transparency from the copy machine. Cut the templets out and using double-sided tape, place the equipment where it belongs. The master (on white bond) can make an unlimited number of copies. Over the years, a collection of masters reduces the time needed to layout the new plant or product line. Templets can also be made from plastic templets purchased from stationary supply stores (see Figure 13-5). These plastic templets have shaped holes cut in the plastic. A pencil then outlines as many shapes as can be placed on an $8\frac{1}{2}'' \times 11''$ sheet of white bond paper. Transparencies are then made. Plastic templets are available for bathrooms, offices, locker rooms, and so on. They were produced for the architectural technique, but are also very useful for the templet technique.

The procedure for laying out a plant using the templet and tape technique is to:

1. place the mylar base on table;
2. layout the external walls;
3. cut out the doors;
4. place I beam;
5. place aisles;
6. place internal walls (minimize);
7. place equipment per flow analysis and activity relationship analysis;
8. using clear plastic overlay, show the flow of material; and
9. seek advice (part of next chapter).

Three-dimensional models (3-D). Three-dimensional model layouts have the big advantage of illustrating any height problems. New 3-D models are being commercially developed every day. The engineering monthly journals are the best source of where to find these models. Three-dimensional models can be placed on a clear plastic sheet with a grid network of 1'' square. Three-dimensional models are nice, but the expense, the difficulty of copying, and the problem of storage space makes them less desirable. The layout procedure for the three-dimensional model technique is the same as that of the templet and tape technique.

The scales for templet and tape, and the 3-D model technique are the same.

One-quarter inch equals one foot is the most popular scale for plant layout followed by $\frac{1}{8}'' = 1'$. Many commercial templets and 3-D models are available with these two scales. Any other scale can be used if you do not need outside materials.

Figure 13-3 Plant and Office Layout Tape

Catalog Number	Tape Surface	Price Code
TL382A	Clear	OO
TL383A	Clear	OO
TL384A	Clear	OO
TL385A	Clear	OO
TL386A	Clear	OO
TL387A	Clear	OO
TL388A	Clear	OO
TL389A	Clear	OO
TL390A	Clear	OO
TL391A	Clear	OO

Catalog Number	Tape Surface	Price Code
TL392A	Clear	OO
TL393A	Clear	OO
TL394A	Clear	OO
TL395A	Clear	OO
TL396A	Clear	OO
TL397A	Clear	OO
TL398A	Clear	OO
TL399A	Clear	OO
TL400A	Clear	OO
TL815A	Clear	OO
TL817A	Clear	OO

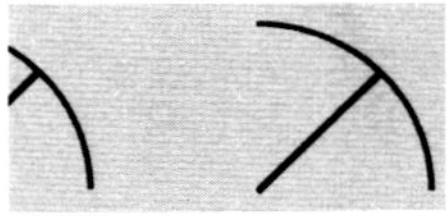

24" Door Swing

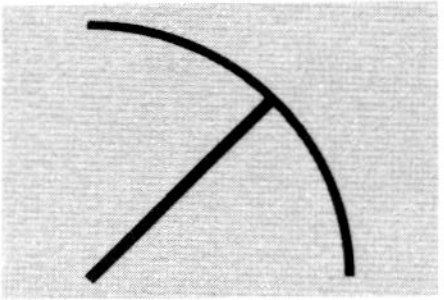

36" Door Swing

Courtesy of Chartpak Co.

Figure 13-3 (cont'd) Plant & Office Layout Tape

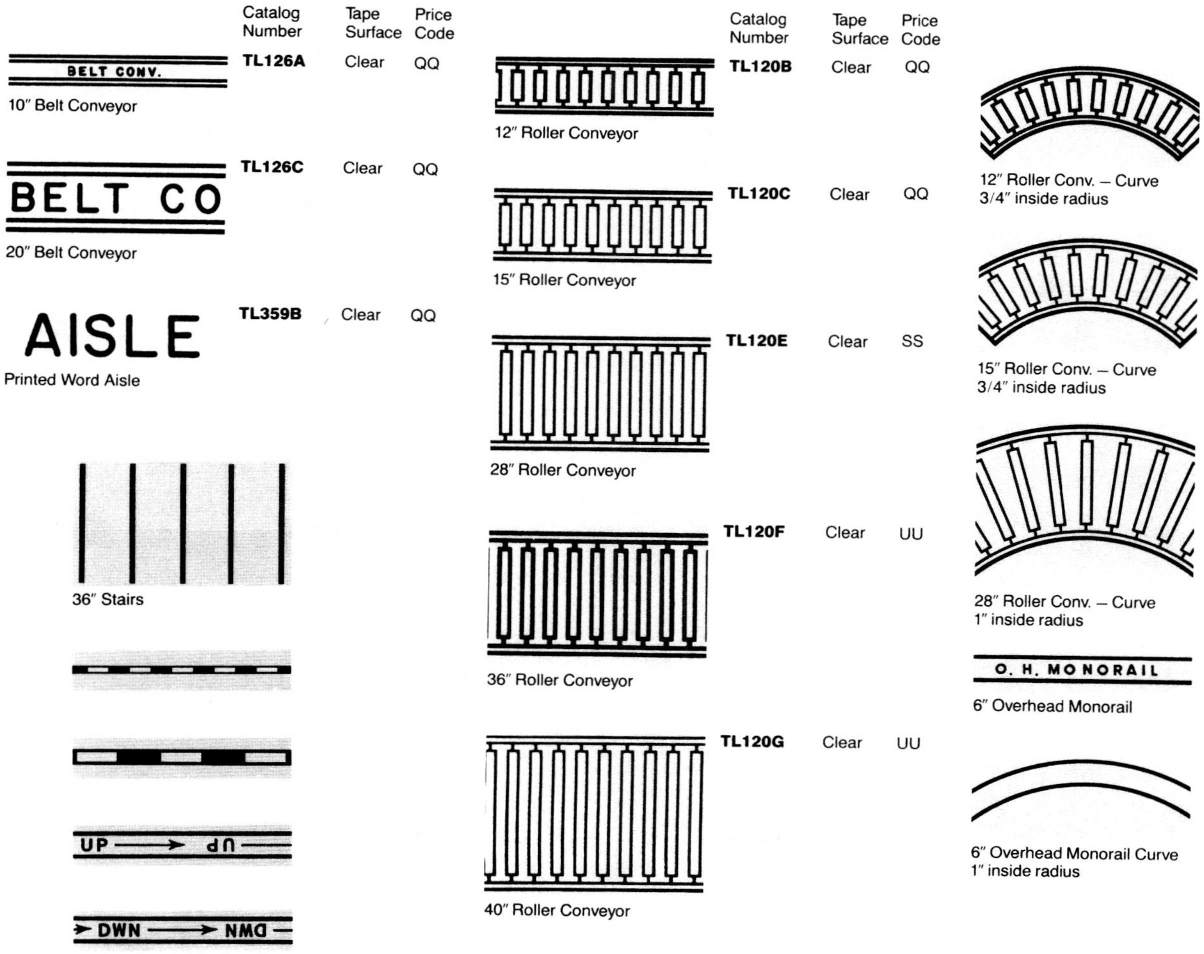

Catalog Number	Tape Surface	Price Code		Catalog Number	Tape Surface	Price Code
TL126A	Clear	QQ		TL120B	Clear	QQ
10" Belt Conveyor				12" Roller Conv. – Curve 3/4" inside radius		
TL126C	Clear	QQ		TL120C	Clear	QQ
20" Belt Conveyor				15" Roller Conv. – Curve 3/4" inside radius		
TL359B	Clear	QQ		TL120E	Clear	SS
Printed Word Aisle				28" Roller Conv. – Curve 1" inside radius		
				TL120F	Clear	UU
				O. H. MONORAIL 6" Overhead Monorail		
				TL120G	Clear	UU
				6" Overhead Monorail Curve 1" inside radius		

12" Roller Conveyor

15" Roller Conveyor

28" Roller Conveyor

36" Roller Conveyor

40" Roller Conveyor

36" Stairs

UP

DWN

Courtesy of Chartpak Co.

MASTER PLAN

Figure 13-4 Sample Plastic Templets

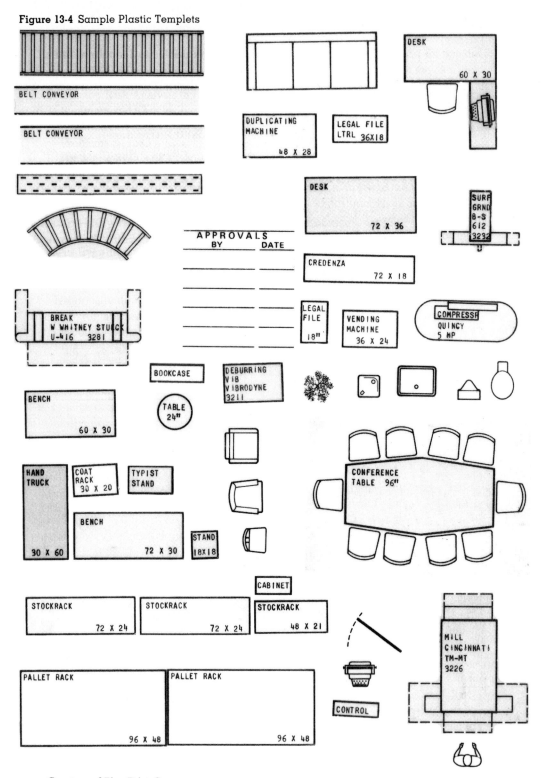

Courtesy of Plan Print Corp.

Figure 13-5 Sample Plastic Templets

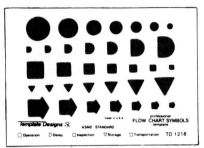

Flow Process Symbols

Contains: standard flow process symbols
— operation, delay, storage, inspection and
transportation — in 7 sizes, 3/16" to 5/8".
ASME. Size: 3 3/4" x 5 1/2" x .030".

No. TD1218

Plumbing — Plan Views

Similar to No. TD1190. Size: 5 1/2" x
9 3/4" x .020".

No. 36T

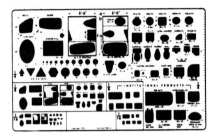

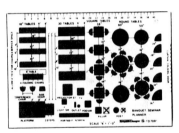

Office Planner

Contains: tables, storage units, L-shaped
desks, bookcases, credenzas, files, platforms
and door swings. Scales: 1/4" and 1/8" = 1'.
Size: 7" x 8 1/4" x .030".

No. TD1529

Courtesy of Alvin and Co., Inc.

Banquet/Seminar Planner

Contains: tables, chairs, door swings,
podium, projector stand, TV, etc. for
planning banquet, seminar or conference
room layouts. Table prespacing is provided
for quick, accurate arrangements. Scale:
1/8" = 1'. Size: 6 1/4" x 4 3/8" x .030".

No. TD7091

A toy company uses $\frac{1}{2}'' = 1'$ scale and is able to write much more information on the templets. A $4' \times 4'$ pallet is reduced to a $2'' \times 2''$ label and the part number, name, and quality per pallet are written right on the templet.

Computer-aided design technique (CAD). Computer-aided plant layout design is the state-of-the-art technique. The advantages of all the previous techniques are improved with CAD and the disadvantages have been minimized. We are assuming a trained CAD operator, the equipment, and the program are available to the company. Of course, this is the only hold up on CAD replacing all previous techniques. New technologists armed with CAD expertise and a knowledge of plant layout will be very valuable to any company.

AUTOCAD was the specific program used to create the drawings in this book, but there are several other techniques. A microcomputer, pen plotter, cathode ray tube (CRT), input device, and a program are needed. With these tools, you can draw on the screen instead of on paper.

The disadvantages of CAD is cost, but once the equipment and program are installed, the cost effectiveness is most important. Changes to layouts can be made very quickly, the quality of drawing are outstanding if plotters are used, and everything that is drawn is kept in the memory system and can be reused many times in the future. The more layouts completed, the easier the job gets. Three-dimensional layouts and layered (overlay) layouts can assist in visualization and spacial relationships. Join the 21st century and use this computer-aided plant layout design technique!

Figure 13-6 compares the four layout techniques to assist in the selection of the best technique for you.

Advanced Computer Systems

Facilities design has undergone a gradual change since the 1940's. It has become more efficient, more useful and (even though more complex) better in every way. From the architectual drawings of the early days, to cut outs of architectual

Figure 13-6 Rating of Layout Techniques

	Architectural	Templet	Model	CAD
1. Skill Needed	High	Medium	Low	High
2. Cost of Equipment	Low	Medium	High	Highest
3. Time to Set Up	Low	Medium	High	High
4. Time to Redraw	Longest	Medium	Medium	Fastest
5. Drawing Time (once established)	Highest	Medium	Medium	Low
6. Changeable	Hard	Easy	Easy	Easy
7. Availability of Material	Good	Moderate	Poor	Moderate
8. Scales Available	Any/All	Few	Very Few	Any/All
9. Selling Tool	Good	Good	Excellent	Excellent
10. Ease of Building File	Low	High	Low	High
11. Ease to Generate Alternatives	Poor	Good	Good	Best
12. Storage Space Required	Moderate	Moderate	High	Low
13. Ability to Copy	Easy	Easy	Difficult	Easy

drawings, to templets, to 3-D models to the CAD systems of today lead us to wonder about the future. I've had a glimpse of this future. I watched a kitchen being layed out using a new computer program. The company's product had been drawn into the database. Starting with the dimensions of the new kitchen, drawer units, cupboards, stoves, refrigerators, sinks, and an island were placed on the grid of the computer screen. Once the top view was completed (Figure 13-7) the auxiliary views were also completed. A view of the kitchen could be had from any one of twelve positions. Figure 13-8 is one of these 12 views. Now, the customer

Figure 13-7 Kitchen Layout

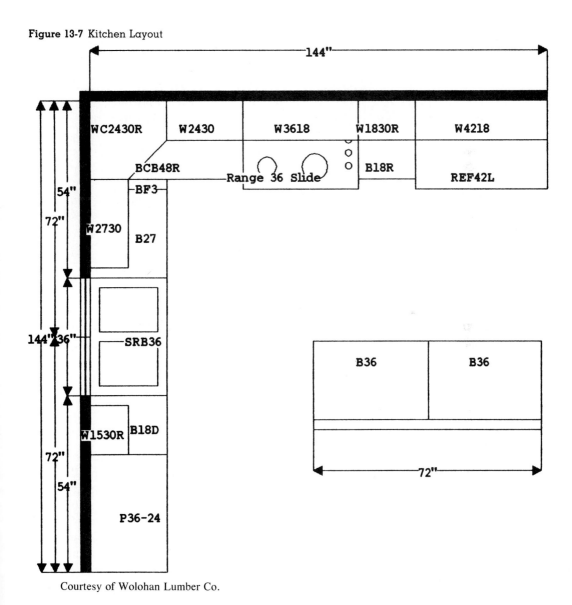

Courtesy of Wolohan Lumber Co.

Figure 13-8 3-D CAD Layout—Kitchen

Courtesy of Wolohan Lumber Co.

can adjust as needed to get exactly what they want. Many people cannot visualize what a facility will look like from a top view layout. These auxiliary views (perspectives) will help all of us visualize the new facility. Any of the auxiliary views can be printed out, and the computer will cost (price) this layout. The 3-D CAD system will be a big boom to facilities design. This will require an extensive database that will require all manufacturers to supply their equipment's 3-D drawings.

Virtual-reality technology will allow the designer (and anyone else) to walk through the plant. It will be almost like being there but there is no plant yet. Virtual-reality is barely out of the laboratory, but a Japanese company (Matsushita Works Ltd.) has kitchen showrooms which allow customers to "walk through" different kitchen designs without leaving their chair. They would wear a special set of goggles hooked up to a computer. A 3-D picture of the newly designed kitchen will appear on the inside of the goggles. An electronic glove allows the wearer to steer themselves through the kitchen-almost like being in your new kitchen. This technology will revolutionize layout. Not only will it assist

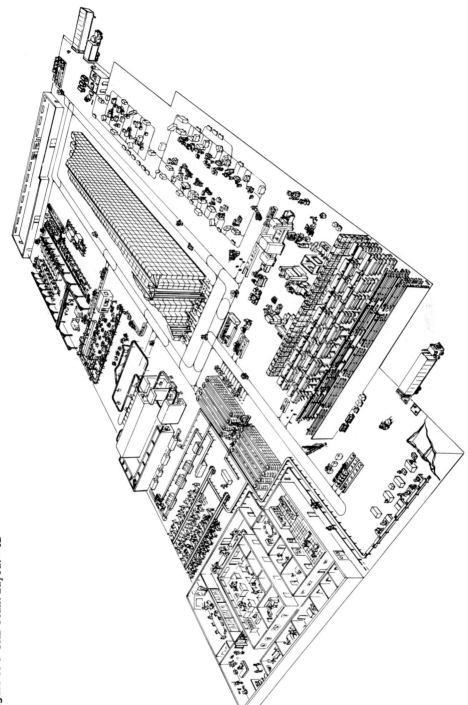

Figure 13-9 CAD Plant Layout—3D

Courtesy of S.I. Handling Systems, Inc.

Figure 13-10 CAD Plant Layout

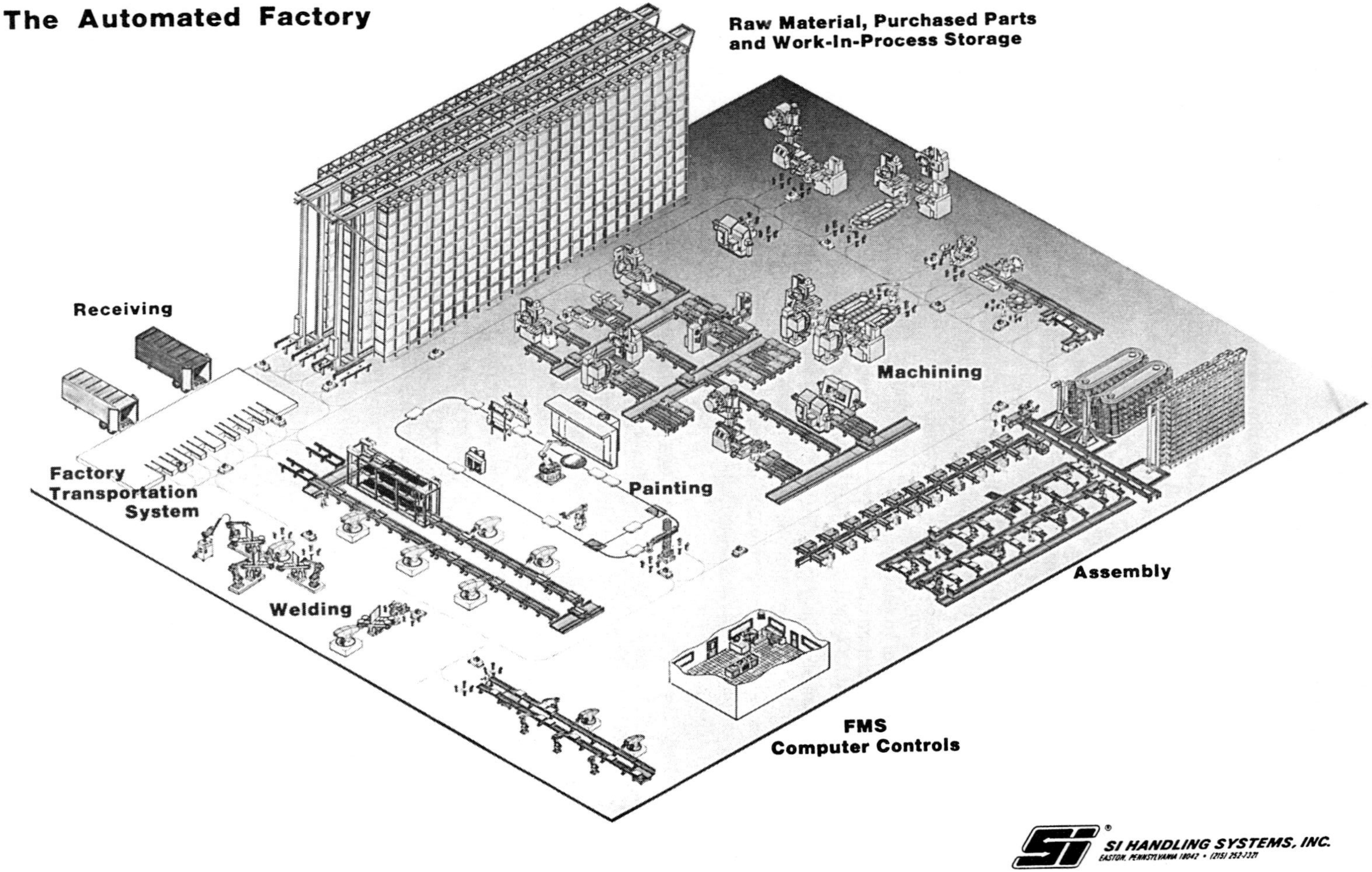

The **Automated Factory**

Raw Material, Purchased Parts and Work-In-Process Storage

Assembly

Machining

Painting

FMS Computer Controls

Welding

Factory Transportation System

Receiving

SI HANDLING SYSTEMS, INC.
EASTON, PENNSYLVANIA 18042 • (215) 252-7321

Courtesy of S.I. Handling Systems, Inc.

the designer, it will help the designer sell their plans. Virtual-reality technology is developing fast, but cost efficient plant layout systems will require years of work, but the direction of plant layout is clearer and the future looks very good.

The future of facilities design is exciting. The computers will become easier to use. The databases will become more compatable interchangable, and more accessable; Process time for new layouts will be speeded up; COSTS of facilities design will become cheaper; the analysis will be faster and better, and the selling of the finished product (our layout) will be easier.

Figures 13-9 and 13-10 are two state of the arts CAD drawings provided by S.I. Handling systems.

PLANT LAYOUT PROCEDURE—TOOL BOX PLANT

This is where everything comes together. The area allocation diagram, as shown in Chapter 12 shows the shape and position of every department and service area. Many of the departments have already been layed out but they now must be fitted to the area allocation diagram considering material flow and size constraints. Some modification may be needed in the area allocation diagram and/or the department layout. The following departments have layouts earlier in the book.

Figure #	Department
3-7	Spotweld
3-8	Assembly and packout
6-9	Workstations (fabrication)
6-10a	Paint department
7-2	Steel receiving
7-3	Parts receiving
7-9	Stores layout
7-7	Shipping
p. 138	Warehouse
7-22	Maintenance
8-3	Employee entrance
8-5	Locker room
8-7	Toilets
8-9	Cafeteria
8-12	Medical Service
p. 266	Office

Using the area allocation diagram as our guide, these layouts must now be coordinated into a final master layout.

The plant layout procedure starts with the exterior walls being located. This is, of course, our constraints. Once the exterior walls have been established, exterior doors, columns, and aisles are located according to the area allocation diagram. Now, one department at a time plus all the equipment and facilities are put in place.

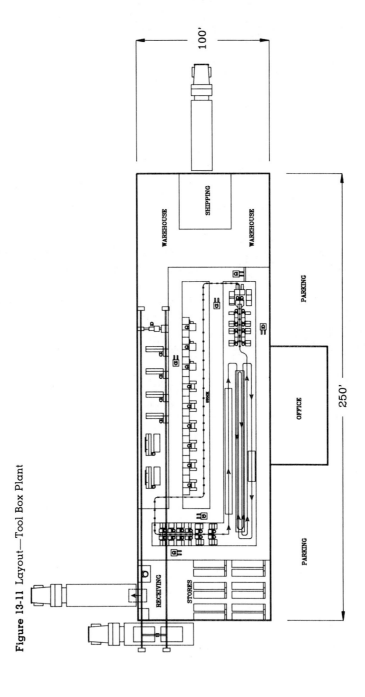

Figure 13-11 Layout—Tool Box Plant

296

The flow of material must always be considered. We considered material and people flow at every step along the process of laying out the plant, but the flow of material out of one department must line up with the starting point of the next department.

The final entry to the plant layout is material space. Everything must have a place otherwise it will be in the aisle. Once everything is in place on the layout, the layout person should follow the flow of every part from receiving to shipping to ensure every requirement has been accounted for. This is the flow diagraming technique shown in Chapter 4. Figure 13-11 is the final layout of our tool box plant. See how it compares to Figure 4-13. Figure 4-13 is the existing layout where Figure 13-11 is our planned improvement. Is it better? Evaluation is needed to answer this question.

OFFICE LAYOUT FOR TOOL BOX PLANT

Starting with the organization chart in Figure 2-6, we can determine that the number of employees in the office is eleven. Eleven people times 200 square feet per person equals 2,200 square feet. The level of the organization technique would require:

Plant Manager	200
Secretary	100
Controller	150
Accountant	75
Production Manager	150
Manufacturing Engineer	100
Supervisors	75
Supervisor	75
Purchasing Manager	150
Plant Engineer	150
Maintenance Supervisor	75
Total:	1,300 square feet
100% allowance	1,300 square feet
Total needed:	2,600 square feet

Between 2,200 and 2,600 square feet are needed.

$$\sqrt{\frac{2,200}{2}} = 33' \quad \sqrt{\frac{2,600}{2}} = 36'$$

Because it's a round number, $35' \times 70'$ will be chosen. $35' \times 70' = 2,450$ square feet.

Figure 13-12 Electric Power Plant Coop Office

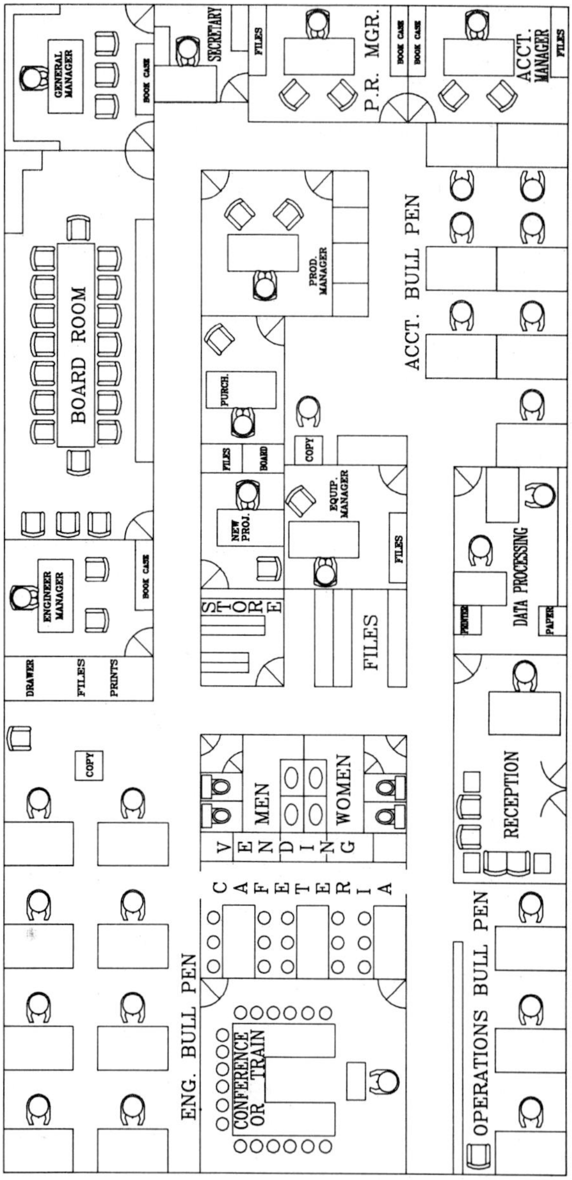

The power plant office layout offers a better example of a detailed office layout. Figure 12-4C is the area allocation and Figure 12-5 is the office area requirements summary. With these two resources, the requirements must be fitted into a 60' × 120' space. Figure 13-12 is the resulting layout.

EVALUATION

When deciding upon which method or which alternative is best, measurements of performance must first be developed. In the beginning of the book, plant layout objectives were established. Did we meet the objectives? Which alternatives met the objectives best? Performance measurement techniques were discussed throughout the book. They are listed here one more time so as to stress their importance.

1. *Minimize distance traveled.* How many feet does a part travel through the plant? The shorter the better. Some travel is not as bad as other methods.

    a) How many feet are traveled automatically? This would be expressed as a percentage but the automatic feet of the total feet is more descriptive and meaningful.

    Example: 1,525 of 2,000 is 76 percent. Seventy-six percent may be graphed (and it should be). One thousand five hundred twenty-five feet of automatic travel out of 2000' shows us how well we have done and how much room for improvement is left.

$$\text{Automatic travel ratio} = \frac{\text{automatic feet}}{\text{total feet}}$$

    b) Gravity movement is free power. If we want to promote the use of gravity, we will calculate the percentage of footage being traveled by use of gravity, and will graph our progress month after month.

$$\text{Gravity ratio} = \frac{\text{gravity feet}}{\text{total feet}}$$

2. *Maximize space utilization.* This can be measured, charted, and improved. We can increase this utilization in many ways. They are measurable in a few ways.

    a) Aisle space can be calculated by the square footage of aisle space and divided by the total space available.

$$a = \%\text{ aisle space} = \frac{3,150 \text{ sq. feet of aisle}}{10,000 \text{ sq. feet of plant}} = 31.5\%$$

    Plot this percentage on a graph and measure it month after month to show improvement. An improvement would be a lower percentage rate.

b) Stores and warehouse cube utilization is total storage available. Length times width times height of your stores and/or warehouse equals the total cubic feet of storage available. We may approach 100 percent, but aisle space, space between materials, and not stacking to the full height creates use in the 30 percent to 40 percent level. It should be our goal to improve our cube utilization. A measure of this is

$$b = \% \text{ of cube utilization} = \frac{\text{storage cubic feet}}{\text{total cubic feet}}$$

c) Machine space utilization is calculated as:

$$c = \text{Machine space utilization} = \frac{\text{machine space required}}{\text{total plant space}}$$

An increase in this percentage would show a lessening of material in process, aisle space and services.

3. A machine may have the capacity to cycle 1,000 to 2,000 times per hour, but an operator must unload, aside, get next part, load and trip the run buttons. This lowers the standard to 250 to 500 per hour (25 percent utilization). An automatic loading can increase the output by 400 percent. To promote this goal:

$$a = \% \text{ Automatic loading machine} = \frac{\text{\# automatic loaded machines}}{\text{total machines}}$$

$$b = \% \text{ Machine utilization} = \frac{\text{time standard}}{\text{theoretical maximum}}$$

b can be for a single machine or the entire department or plant. We want to approach 100 percent.

4. Control material handling costs.

a)
$$\% \text{ Material handlers} = \frac{\text{\# material handlers}}{\text{\# production people}}$$

or

$$\text{Material handling ratio} = \frac{\text{\# material handling hours}}{\text{total hours worked}}$$

b) Manual move/operations ratio equals the number of moves divided by the number of operations. This will promote the combining of operations or the mechanization of moves to eliminate manual moves.

5. Just in Time manufacturing ratios measure how long a product is in process (in the plant). We want to move material through the plant as fast as possible to reduce inventory and inventory carrying costs. If we add up all the time standards in hours per unit, we would have the theoretical shortest time a product

would be in a plant. An appliance manufacturers makes nearly every part and assembles the washing machine in $3\frac{1}{2}$ hours or less, yet they have millions of dollars worth of inventory in the plant. If the inventory dollar value is divided into the total annual sales, we get the number of turns a year (turnover of inventory). Two turns a year equals 6 month's worth of inventory. Three and one-half hours divided by 2,000 hours (6 months with 2 shifts per day) is a very small percentage. Through the use of many of the discussed techniques, this

Figure 13-13 Plant Layout Ratio Progress Graphs

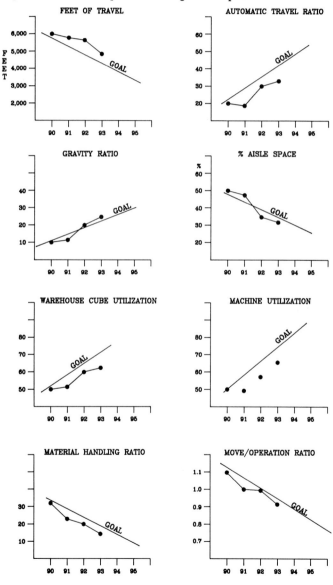

percentage can be increased to over 10 percent. The cost savings will be fantastic!

$$\text{In-process time ratio} = \frac{\text{cycle time (total)}}{\text{total time in process}}$$

6. The from/to chart is an evaluation technique that is quantitative and results in usable measurable efficiency. The from/to chart is a good example of why measurement and evaluation techniques are so valuable.

7. The cost evaluation technique is the most complete and most used evaluation technique. The total cost of the project, the operative costs, the sales price and the forecasted sales must all be determined with great accuracy and a return on investment (R/I) calculated. This results in budgets and operations plans which result in the profit goals of the company. The cost evaluation technique is mandatory for new plants, and good management for continuing operations.

All the above measurements can be evaluated on a continuing basis and charted. Figure 13-13 shows an example of the use of ratios and key indicators of layout efficiency improvement.

QUESTIONS

1. The layout is only as good as what?
2. What are the two types of layouts?
3. What flow analysis technique depends upon the plant layout?
4. What is a plot plan?
5. Which is more costly, a front foot or an additional foot of depth?
6. Where will we expand to? Factory_____ Office_____
7. How much property should we buy?
8. What is the master plan?
9. What are the four methods of constructing a master plan?
10. Which is the most expensive? Why?
11. What are the most common scales? List two.
12. Is there a time when the architectural technique would be best?
13. What are 10 measurements of performance used in evaluating layout alternatives?
14. What should the trends be for the following?
    a) distance traveled
    b) automatic feet ratio and gravity feet ratio
    c) aisle space

d) cube utilization

e) machine space utilization

f) percent automatic loading

g) machine utilization

h) percent material handlers

i) material handling ratio

j) in process time

Chapter 14

Selling the Layout

The easy part is over. Now is the time to seek the approval of your months of work. This entire book has been devoted to collecting and analyzing the data to produce the best layout possible. If management can follow our reasoning, they will come to the same decision we did. Our job in "selling the layout" is to lead them through our reasoning process. The written project report should do exactly that: Lead the reader to your conclusion. The biggest mistake made by layout technicians is assuming management knows more than they do about the project. Assume they know nothing (just like you when you started this project) and show them the systematic approach you took.

THE PROJECT REPORT

The project report outline was introduced in Chapter 1 in the 24-step plant layout procedure. Now that we have layed out the tool box plant, the following could be the specific outline for the project report:

1. The goal is to layout a manufacturing plant and support services in order to produce 2,000 tool boxes per eight-hour shift and to achieve the subgoals of:
   a) minimizing unit cost;
   b) optimizing quality;
   c) promoting the effective use of people, equipment, space, and energy;
   d) providing for the employees convenience, safety, and comfort;

e) controlling project cost;

f) achieving the production start date of 12-1-93; and

g) minimizing work in progress inventory.

2. Set a volume and plant rate (R value):

a) 2,000 units per day;

b) 10 percent personal fatigue and delay allowance;

c) 80 percent historical performance; and

d) an R value of .173 or 5.8 sets of parts/minute from every operation in the plant.

3. Drawings of the product should include:

a) blueprints (Figure 2-1);

b) an assembly drawing (Figure 2-2);

c) an exploded drawing (Figure 2-3); and

d) a parts list (Figure 2-4).

4. Set a management policy. It should include:

a) an inventory policy—maintain a 30-day supply;

b) an investment policy—50 percent ROI;

c) a start-up schedule—date;

d) a make or buy decision (Figure 2-5); and

e) an organization chart (Figure 2-6).

5. The process design should include:

a) a route sheet for each "make" part (Figure 3-1 and 3-2) including time standards;

b) the number of machines required (Figure 3-3);

c) the assembly chart (Figure 3-4);

d) assembly time standards (Figure 3-5);

e) conveyor speeds (paint 17.34 ft./min., assembly 11.56 ft./min.);

f) assembly line balance (Figure 3-6);

g) the subassembly line layout (Figure 3-7);

h) the assembly and P.O. layout (Figure 3-8);

i) the process chart (Figure 4-11);

j) the flow diagram (Figure 4-13);

k) the operations chart (Figure 4-14); and

l) the flow process chart (Figure 4-15).

6. The activity relationship should include:

a) the activity relationship diagram (Figure 5-1);

b) the worksheet (Figure 5-2);

c) the dimensionless block diagram (Figure 5-3);

d) the flow analysis (Figure 5-3).

7. The workstation design should include:

a) machine layouts (Figure 6-4 to Figure 6-8);

b) area determination (Chapter 6);

c) paint layout (Figure 6-10a); and

d) aisles.

8. Auxiliary services should include:

a) receiving (Figures 7-2 and 7-3);

b) shipping (Figure 7-7);

c) stores (Figure 7-15);

d) warehouse (Figure 7-24);

e) maintenance (Chapter 7);

9. Employee services should include:

a) parking lots (Chapter 8);

b) employee entrances (Figure 8-3);

c) locker rooms (chapter 8);

d) toilets (Chapter 8);

e) the cafeteria (Figure 8-9); and

f) medical services (Figure 8-12).

10. The office should include:

a) an organization chart (Figure 2-6).

11. Area allocation should include:

a) the total space requirements worksheet (Figure 12-1);

b) the building size (Chapter 12);

c) the dimensionless block diagram (Figure 12-2); and

d) the area allocation diagram (Figure 12-3).

12. The layout should include:

a) the plot plan (Figure 13-2).

b) the master plan (Figure 13-11).

THE PRESENTATION

The presentation of the project occurs at a management meeting where the project engineer (or engineers) present their project plan. The presentation should be visual. Otherwise, the managers could read the report and there would be no need for a meeting. The two most visual items are the product model and the layout.

Using the product model, the presenter can cover:

1. The goal and subgoals.
2. The volume and plant rate.
3. The product.
4. The make or buy decisions.
5. The process design

With the layout, the presenter can cover:

1. The process design (further description on flow of each part).
2. Assembly and packout.
3. The operations chart or flow process chart.
4. The activity relationships and dimensionless block diagrams.
5. The auxiliary services.
6. The employee services.
7. The office.
8. The area allocation diagram.

The plot plan will show how the plant sets on the lot. A real presentation would present a cost budget, but that subject would take another book, so we won't cover it here.

ADJUSTMENTS

The plant layout technician should present the layout to every person who will listen. Your friends will criticize your project to help prevent costly errors, your enemies will tell you "great job, go get them" (meaning take it to management and make a fool of yourself). With every presentation, you will adjust the layout making it better and better.

APPROVAL

Once you have a completed project (the schedule probably dictated the date), a formal presentation (or presentations) are required. The first presentation would be to your supervisor and the production manager. Their great experience will almost always point out problems with your plan. Depending on the magnitude of the problems, they may sign off (approve) the project subject to the suggested changes.

More major changes may require a second presentation. Most companies will require many levels of approval depending upon the amount of money being requested.

At one company, I presented the layout to the general manager of the plant and he approved. He did not have the authority to approve the $75,000 expenses, however, so we flew to Los Angeles and presented the proposal to our division president and he approved. He did not have the authority to approve that much money, so he asked to meet us the next week in New York. In New York, the project was approved.

The approval process is important, and those top managers did not get there without a lot of experience. Their input is valuable and will only serve to make a better project. When the project is successful, you will get a lot of credit because you made it happen. Whenever a top manager makes a suggestion that you incorporate, you make them a part of the project and you have recruited another person who has an interest in the success of your project. We want to involve everyone to ensure their cooperation. What management is approving most of all is an expenditure budget (limit). Successful project engineers and managers will never exceed the budget.

THE REST OF THE PROJECT

This book is almost at an end, but it would be a mistake not to mention a few very important topics not covered as of yet.

Sourcing

Sourcing is the process of finding suppliers who can provide the equipment, materials, and supplies needed for a project. These suppliers can be very helpful to the project engineer. Not only can they provide information on exact models, speeds, feeds, cycle times, and cost, they will help with special design requirements, with savings calculations, and even do some of the layout work. It is quite normal to work with three suppliers of each piece of equipment, but these suppliers expect to get some of the jobs—not every one, every time. If a supplier feels you are using him/her, they will be unwilling to help in the future. The results of sourcing is a list of equipment and supplies needed to create the layout you designed, and a specific source and price. The total of this list dollars is a major part of your project budget. The day the budget is approved, you can spend 70 to 80 percent of the dollars because you have chosen the supplier and had a purchase order waiting for approval.

The purchasing department normally does all the company's buying, but sometimes (especially in building a new plant) the purchasing function is delegated to a project manager. This project manager is totally responsible for getting the job done and within budget. Either way, the purchasing department should be involved because of their special skills and knowledge. If the project manager works through purchasing, the purchasing person will want to know your desires and needs, and will appreciate the help. Still, on the day the budget is approved, the purchase orders are released and equipment materials and supplies start moving toward our sight.

Installation

Once the new plant is built or the existing plant is readied, the equipment starts arriving. This equipment needs to be placed and connected to power, water, and/ or air. The delivery time varies from purhcase to purchase, and some special pieces of equipment can take months to arrive. Once the equipment enters the plant, *installation* can also take months. A chrome plating machine or powder paint system are good examples of this process. The installation costs money so it must be part of the budget. The installation takes time and must be a part of the schedule. The installation ends with the project engineer (or his/her designated alternate) trying out the machine. This may also be an engineer from the supplier who is training our staff, but it must be proved before production begins.

Engineering Pilot

An *engineering pilot* is a tryout of all the tools, equipment, and raw materials to see if we can make a product. At least one of every workstation must be available. The first small order of parts or raw materials must be available, and a few production people are asked to run every operation. There are always problems starting up anything new, and the engineering pilot lets us find the problems with machines, tools, and materials very soon so they can be corrected. The results of an engineering pilot may be a few new products, but mostly, a list of problems that must be fixed before production begins.

The product engineers (designers of parts), the purchasing management (the providers of raw material and finished parts), the quality control engineers (to anticipate quality problems), the tooling engineers (the designers of tooling), the industrial engineers (the designer of workstations and standards) and the facilities design project manager (the boss) all want to be a part of the engineering pilot. After the pilot, a meeting is conducted where all the problems are reviewed, discussed, and assigned. This must be a very closely knit group of people.

Production Start

Within two weeks to a month of an engineering pilot, *production* will start. This is the worst day of a layout person's life. Everything up until this point has been fun. Seeing the plan come together is great, but when the production people show up in mass wanting to go to work, you, the supervisor and lead person must train everyone. Everything is supposed to work as planned, but it does not, so you need to direct maintenance work, get parts reworked so they will fit, adjust machines, retrain people, and most important of all, make a list of what needs to be fixed before tomorrow morning. When the people go home at the end of the shift, your day is about half over. You need to get everything fixed by tomorrow morning. This is a hectic time and most project engineers feel they are the most productive during production's start.

Production efficiency for a second-year product averages 85 percent in a plant with a performance control system. First-year products average about 70 percent for the entire year which means, in the beginning of the production year, performance may be as low as 50 percent or less. This low performance is normal (you will experience this) and must be anticipated in order to maintain the delivery schedule. This low performance increases costs as well, and must be a part of the start up budget. A part of this cost is recognized by using the first year efficiency of 70 percent when calculating the R value (plant rate).

Debugging and Follow-Up

Debugging is a common term used to describe the process of making the plan work. Getting the bugs out of every operation in order to perform properly. Depending upon the complexity of the product and the processes, debugging can last from two months to a year. After the debugging period is the *follow-up* period. The dividing line between debugging and follow-up is invisible and there is no ending to the follow-up. Once we stop following up, improvement stops and the product starts to die.

Conclusion

The plant layout procedure described in the first section of this chapter is a good outline for most plant layout projects. Not every step is used in every project, but skipping a step must be done after thoughtful consideration. The tool box plant didn't need a from/to chart because every part flowed through the same sequence of machines. The results were obviously 100 percent, so why bother? This is an example of thoughtful consideration to eliminate a step.

Plant layout projects are mostly fun. The greatest influence you will ever have on a plant's effectiveness and efficiency (doing right things right) is laying out a new plant. A relayout is second. Industrial managers do not give projects that are too big for project engineers. Project engineers have to prove themselves on small projects before they earn the right to work on a big project. A project engineer almost needs to beg for the big projects, so don't worry about being given a big project too soon, the other problem (needing too big) is more often the truth. Accept every project that is offered to you with enthusiasm and do a great job on the small jobs and you will earn the big jobs sooner than you can imagine. Have courage! If you've diligently studied this book, you are ready.

Answers

Chapter 1

1. Plant layout is the organization of the company's physical facilities to promote the efficient utilization of equipment, material, people, and energy.

2. Facilities design includes plant location, building design, plant layout and material handling.

3. Material handling is defined simply as moving materials.

4. Ask these four questions in this order:

   a) Can I *eliminate* this cost?

   b) Can I *combine* costs?

   c) Can I *rearrange*?

   d) Can I simplify?

5. a) 50% b) 40–80%

6. a) Minimize unit cost.

   b) Optimize quality.

   Promote the efficient utilization of:

   c) people;

   d) equipment;

   e) space;

   f) energy.

Provide for employees:

   g) convenience;

   h) safety;

   i) comfort;

   j) control project costs;

   k) achieve production start date.

7. A mission statement is a simple statement of quantity, quality product, and cost goals used to keep our minds on track.

8. Item 12 and 13 product #1670.

9. This approach is a systematic approach that results (like magic) in a great plant layout.

10. The 24 steps (Figure 1-2).

11. New plant, new product, design changes, and cost reduction.

Chapter 2

1. Marketing, product design, and management policy.

2. Selling price, sales volume, seasonality, replacement parts.

3. The plant rate in decimal minutes (how fast we must produce every part).

4. Working minutes, efficiency history, down time and number of units to produce.

5. It determines the speed of the entire plant.

6. Blueprints, parts list or bill of materials, model shop samples.

7. Investment policy, inventory policy, start up schedule, make or buy decision, organization chart, feasibility studies.

8. Figure 2-5 a list of parts that we will make and a list of parts that we will buy.

9. Purchasing because they will buy the part if it's cheaper on the outside.

Chapter 3

1. How are you going to make each part, what equipment, what time standard, tools, sequence of assembly, etc.

2. Fabrication and assembly/packout.

3. A sequence of operations to make a part.

4. Part number, part name, quantity to produce, operation numbers, operation description, machine numbers, machine games, tooling needed and time standard.

5. How many units per day are needed, what machine runs what parts, and what is the time standard for each operation.

6. Decimal minute.

7. The assembly chart shows the sequence of operations in putting the product together.

8. Number of units needed per minute times the distance between leading edge of one unit and the next unit.

9. Hook spacing and parts per hook.

10. Equalize work, identify bottlenecks, establish line speed, determine number of work stations, determine product cost, establish percent load of each person, assist in layout, and reduce production costs.

11. a) 0.6621 b) 3104 c) 25.44 d) A1

   e) Yes, because the total hours per unit is less.

   f) $23,030 per year.

Line Balance Improvement

Operation	Time Standard	# Stations	AVE Time	% Load	Hrs./Unit	Units/Hr.
SSSA1	.306	2	.153	99	.00517	194
SSA1	.291	2	.146	94	.00517	194
SSA2	.260	2	.130	84	.00517	194
SA1	.356	3	.119	77	.00775	129
A1	.310	2	.155	100	.00517	194
A2	.555	4	.139	90	.01033	77
A3	.250	2	.125	81	.00517	194
S2A	.415	3	.138	89	.00775	129
S3A	.250	1.44	.250	Sub	.00417	240
P.O.	.501	4	.125	81	.01033	77
		25.44			.06621	

12. Mass production and job shop.

13.

Operation #	Time Standard	Actual # Stations	Average Stations Cycle	% Load	Hours Per Unit	Pieces Per Hour
1	.390	2	.195	78	.00834	120
2	.235	1	.255	94	.00417	240
3	.700	3	.233	93	.01251	80
4	1.000	4	.250	100	.01668	60
5	.240	1	.240	96	.00417	240
6	.490	2	.245	98	.00834	120

Chapter 4

1. The path a part takes through a plant.

2. Minimize distance travelled, backtracking, cross-traffic, and cost.

3. Fabrication and total plant.

4. String diagram, multicolumn process chart, from-to chart and process chart.

5. 65 percent. See Attachment A.

Attachment A

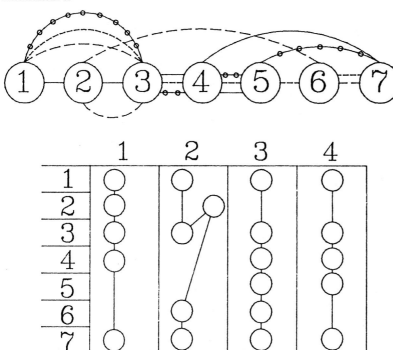

	1	2	3	4	5	6	7	N.P.	P.P.
1		①1	⑱2+3+4					10	19
2			①1			⑧2		3	9
3		④2		8 1+3+4				10	12
4					⑦3+4		③1	8	10
5						③3	⑧4	7	11
6							⑤2+3	5	5
7								43	66

6. Using a standard form, show all the operations from the route sheet (Figure 3-1) add transportation. Taken distances from flow diagram Figure 4-13, if $1'' = 40'$. Place inspections, delays and storages where they belong.

7. Flow diagram, operations chart and flow process chart.

8. Attachment B (see Figure 4-14c).

9. Operations chart and the process charts.

Chapter 5

1. A = Absolutely necessary that these two departments be close.
E = Especially important.
I = Important
O = Ordinary importance.
U = Unimportant.
X = Undesirable.

2. Your talks with management.

3. A reminder of why you coded something the way you did.

4. Compare with class. Look for the best answer, the one with the least checkmarks.

Chapter 6

1. Anywhere, because you will always make improvements so the first sketch will be wrong.

2. The cheapest way to get into production because any additional expense must be cost justified.

3. See Chapter 6, page 84.

4. Guidelines for efficient and effective workstation designs.

5. Doing right things.

6. Doing things right.

7. Aisles, work in progress and small miscellaneous extra room.

Chapter 7

1. Receiving, stores, warehouse, shipping, maintenance, tool room and utilities.

2. Similar people, equipment and space requirements.

3. Common equipment personnel, spaces are used and reduced facility costs.

4. Space congestion and material flow.

5. No! What is most efficient.

6. Morning delivery and afternoon pickup is standard,

7. Less than truck load (common carrier business) uses break bulk stations.

8. Only one (or a few) trucks would show up in the morning instead of many trucks showing up around the clock.

9. See Chapter 7, page 103.

10. A sequential numbering stamp and system to record the order of receiving.

11. The actual day number of the year. January 1st being number 1 and July 1st being number 183.

12. An over shortage or damage report made out by the receiving clerk sent to purchasing for resolution.

13. A notice to the rest of the company that a product has been received.

14. Depends upon arrival rate (trucks per hour) at peak, and service rate (unloading time).

15. How many trucks must be serviced per hour?

16. Parking space, maneuvering space and roadway.

17. Comparison of the pounds of finished product produced in one day to the size of a semi trailer that holds 40,000 pounds. If we produced 100,000 pounds of product per day, $2\frac{1}{2}$ trailers could bring the raw material into our plant.

18. Listed in Chapter 7.

19. Trucking companies charge by the pound, a quality control technique. To load trucks, a measure of efficiency.

20. The truckers authorization to remove product from our plant and part of the trucking company's billing process.

21. A place to hold raw material and supplies.

22. Raw material, finished parts, office supply, maintenance supply, janitorial supply, etc.

23. Size of parts, number to be stored.

24. "A" Inventory items are the 20 percent of the parts that account for 80 percent of the value of material. "B" items are the 20 percent of the parts that account for 15 percent of the value and "C" items are the 60 percent of the parts that account for 5 percent of the value. If we can reduce the

6, 7 1, 3, 5, 14, 11, 12,	5, 7, 6 3, 14, 11, 12,	4, 5, 15 1, 2, 3, 11, 10, 14, 16 13,	
5, 2, 16 8, 10, 11, 14, 13,	4, 6, 15, 16 2, 3, 5 7, 8, 11, 14,	3, 5, 15, 4 1, 2, 12, 8, 11, 14,	13 8, 3, 10, 14, 12, 15, 16

$$X = 4 - 2 \; \checkmark$$
$$2 - 4 \; \checkmark$$
$$A = 15 - 5 \; \checkmark$$
$$5 - 15 \; \checkmark$$
$$3 - 9 \; \checkmark$$
$$9 - 3 \; \checkmark$$

—————————

$$6 \; \checkmark$$

1, 3, 5, 16 2 4, 15, 10, 14, 11,	2, 4, 8, 9, 10, 1, 5, 3 12, 15, 6, 7, 11, 13, 14,	3, 9, 10, 8 4, 5, 14, 16 13,	

2,   3,
 	
1
4, 7, 15,
    10,
14,   11.
	3, 8, 9,
	    11,
	10
	1,
	2, 13, 14,
	12, 15, 16
		3, 8, 10,
		
		9
		
		14,	

11,
 	
12
3, 4,
    6,
10, 13,   7,
	10, 12,
	14,
	11
	3,
	4, 5, 1, 2,
	15, 16 6, 7,
		11,
		
		14
		    10,
		1, 2, 3, 13,
		4, 5, 6, 15,
		7, 8, 9, 16	

stores inventory of "A" items, we can reduce the space requirements and inventory carrying costs.

25. About 25 percent of the value of the inventory each year and includes interest, taxes, insurance, space, utilities, damage and obsolescence.

26. Just-in-time is a new inventory policy that stresses having only enough inventory to run a few hours.

27. Maximize the utilization of the cubic space, provide for immediate access to everything, and provide for the safekeeping of the inventory.

28. Review Figure 7-8.

29. The inventory curve (Figure 7-8) shows that on the average only one half of the inventory is on hand (it's full on the first day of receipt and empty on the last day), so when material come, in, it is placed anywhere an empty spot exists.

30. Every location in the storeroom is identified with a location number, when something is placed in this location, the location number is recorded in the locator system.

31. One foot of aisle access, on both sides of the aisle, we have two aisle feet of access with every foot of length of the aisle.

32. See layout in Attachment C.

33. For storage of finished product.

34. Fixed locations and a small amount of everything.

35. Safekeeping of finished goods and maintain some stock of every product our company sells.

36. A function of the warehouse that collects a customers ordered goods.

37. By identifying the most popular items sold by our company and locating those items more conveniently.

38. The layout of cartons on a pallet to insure safe loads and maximum cube utilization.

39. A form of balcony built over an area to use the space above our heads.

40. From 2 to 4 percent of the plant's personnel.

Chapter 8

1. See Chapter 8, page 141.

2. Depends on the number of employees and the employees to parking space ratio.

3. 250 square feet.

4. Security, time cards, bulletin boards.

5. Parking, locker room, restroom, and cafeteria.

6. Gives employees space for street clothes, work clothes, personal things like lunches, coats, etc.

7. Four square feet per employee.

8. No further than 500 feet from any employee. No fewer than two (one men's; one women's).

9. The building code will tell you.

10. Normally 15 square feet is required for toilets, wash basins, and doorways and 50 percent extra for aisle. A good rule of thumb is 60 square feet per toilet, and this would include everything.

11. Cafeteria, vending machine, mobile vendors, dining rooms, off-site.

12. Activity relationship diagram plus an outside wall close to restrooms and locker rooms.

13. 10 square feet per employee.

14. Located within 200' of every person.

15. 15 square feet each—including drinking space.

16. As little as possible, but 25 percent would be outstanding.

17. 500 employees equal one nurse.

18. From 36 square feet to 300 square feet per nurse per shift. Three nurses could occupy one 300 square feet area if one was assigned to each shift.

Chapter 9

1. M.H. is the function of moving the right material to the right place at the right time in the right amount in the sequence and in the right position to minimize costs.

2. Part numbering, location, inventory control, standardization lot size, order quantity, safety stock, labeling, and automatic identification techniques.

3. The one that produces the lowest unit cost.

4. See "Goals of Material Handling."

5. The College Industrial Committee on Material Handling Education sponsored by the Material Handling Institute.

6. A summary of generations of experience in material handling engineering. A guideline for the application of sound judgment.

7. See Figure 9-3.

8. See "The Material Handling Problem Solving Procedure."

Attachment C

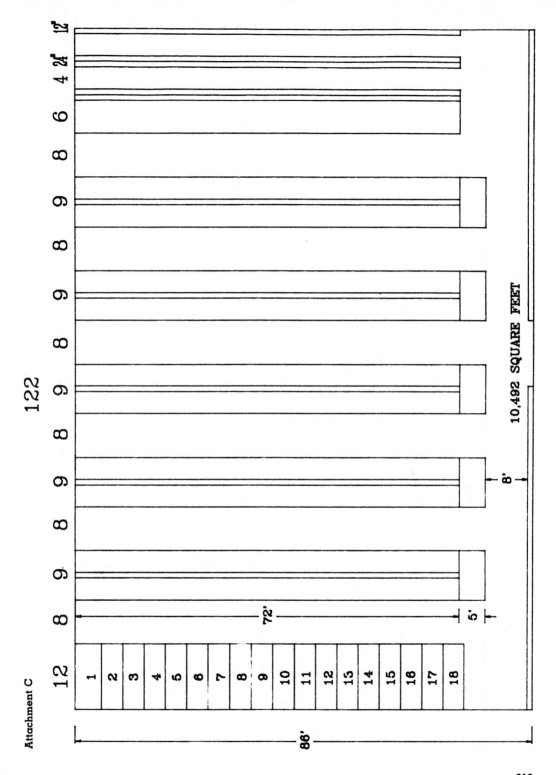

319

Chapter 11

1. See Chapter 11, page 244.

2. Supervisor's, open space, conventional, modern.

3. Easy communication, common equipment, less space needed, easier to heat and cool, easier to supervise, easier to change, common files and literature, easier to clean and maintain.

4. Lack of privacy, noise, status, confidentiality.

5. Privacy, point of use storage, second floor, centralized or decentralized, flexibility, conference room, library, reception area, telephone system, copy machine, mail, file storage, word processing, aisles, computers, lighting, vaults, expansion.

6. Organization chart, procedures diagram, force diagram, activity relationship chart worksheet, dimensionless block diagram, size, and master plan.

7. It would tell us the number of people and their level in the organization.

8. 200 square feet per employee.

9. See Figure 11-11.

10. Tells who talks and works with whom.

11. Circles and lines. 4 lines = absolutely necessary that these two departments (or people) be close together. 3 lines = especially important, 2 lines = important, 1 line = ordinary importance.

12. The force diagram.

Chapter 12

1. Dividing the building's space among the departments.

2. See Figure 12-1. A collection of space requirements for every department and service area to develop the total plant space needs.

3. Under floor, overhead, floor, in the trusses and on the roof.

4. Using the golden rule of architecture, a building is most efficient if it is twice as long as wide. This is a 2 to 1 ratio.

$$\sqrt{\frac{\text{square footage}}{2}}$$ and round off to the nearest column space. This dimension will be the width, and twice this dimension will be the length.

5. See Chapter 12, "Four Step Allocation Procedure."

6. An area allocation diagram, see Figure 12-3c.

7. Mezzanines, racks, shelves, overhead conveyors, etc.

   a. Locker rooms because they are used less often.

   b. Accounting because they have fewer guests.

   c. Old files because they are used less often.

9. Columns are posts that hold up the roof.

10. The distance between columns

11. a. $650' \times 1,300'$ b. $350' \times 700'^*$ c. $200' \times 440'$

12. The dimensionless block diagram.

Chapter 13

1. As the data backing it up.

2. Plot plan and master plan.

3. The flow diagram is drawn on a plant layout.

4. Shows how the building, parking, driveways fit on the property.

5. A front foot.

6. To the *back* of the plant and *up* in the office.

7. Ten times the building space needed.

8. The finished product of a plant layout project. Shows where every maching, workstation, department, desk, etc. are located.

9. Architectural drawing, templet and tape, 3-D and CAD.

10. Architectural because of redraw time.

11. $\frac{1}{4}'' = 1'$; 2nd $\frac{1}{8}'' = 1'$

12. Yes, when it is a one time project.

13. Distance traveled, automatic travel ratio, gravity ratio cube utilization, aisle space ratio, machine space ratio utilization, automatic machine loading ratio, machine utilization, material handling costs ratio, in-process time.

* May be one bay longer or wider.

14. a. down

    b. up

    c. down

    d. up

    e. up

    f. up

    g. up

    h. down

    i. down

    j. down

Index

Material handling:
  checklist, 171–74
  definition of, 2, 157
  equation, 166
  equipment, *see* Equipment, material
    handling
  goals of, 3–7
  importance of, 1–2
  introduction to, 1–9
  problem solving procedure, 170–71
  problems with, 157–74
    20 principles of, 160–70
    checklist, 171–74
    cost justification, 158–59
    goals of, 159–60
  relationship to plant layout, 2
Maximum usage, as part of inventory
  graph, 115
Mechanization principle, material
  handling and, 165
Medical facilities, space requirements
  for, 153
Mezzanines, 135, 192
  definition of, 135
  as type of storage unit, 192
Minimum usage, as part of inventory
  graph, 115
Mission statement, 3–7
  developing a, 3–7
  supporting objectives of, 3
Mobile equipment, stores, 194–99. *See
  also* Stores, mobile equipment
Mobile fabrication equipment, 209–15
  auger conveyor, 211
  chutes, 209
  monorail trolley, 211
  powered hand trucks, 215
  roller conveyor, 209
  skatewheel conveyor, 209
  slides, 209
  vibratory conveyor, 211
Mobile vendors, 148–51
Model shop samples, 18
Modern office concept, 248
Modern office, designing, 248–51
Monorail trolley, 211
Motion economy, 83, 89–90, 94–95
  definition of, 83
  principles of, 89–90, 94–95
    basic motion types, 90
    hand motions, 89
    location of parts and tools, 90
    operator considerations, 94–95
    release the hands, 90
    use of gravity, 94
Motion types, as principle of motion
  economy, 90
Moving equipment, 179–85
Multi-column process chart, 48–50
  sample, 49

N

New plants, as type of plant layout
  project, 8
New product, as type of plant layout
  project, 8
Noise, reducing office, 244

Normal distribution, as part of inventory graph, 115
Normal usage, as part of inventory
  graph, 115
Number of stations, definition of, 39

O

Obsolescence principle, material
  handling and, 169
Off-site diners, 148–51
Offices:
  area allocation, 277–80
  employee parking, 142
  equipment, layout for, 244
  layout, 243–69
    design goals, 243–44
    space requirements, 243–69
    special requirements, 251–56
    techniques, 257–69
      activity relationship diagram, 263
      activity worksheet, 264
      communications force diagram,
        261
      dimensionless block diagram,
        264–66
      master layout, 267–69
      organizational chart, 257–59
      procedures diagram, 259–60
      techniques, site development,
        266–67
  types of space, 245
Office space, types of, 245–51
  conventional, 248
  modern, 248
  open, 245–48
  supervisor, 245
Office supplies stores, 112
One hundred percent station, 39
Open office space, 245–48
Open shipment, as function of receiv-
  ing department, 104
Operating costs, material handling and,
  2–3
Operations chart, 62–65
  design, 64
  procedure for developing, 62, 65
  sample, 63
Operator considerations, as principle
  of motion economy, 94–95
Operator space, definition of, 95
Order quantity, as part of inventory
  graph, 115
Orders, collecting, as function of
  shipping department, 109
Organizational chart, 22, 257–59, 277
  sample, 22
Organizational relationships, 19, 20
Outside areas, dock, 105
Over Shortage and Damage reports
  (OS&D), 104
Overhead space, planning, 273
Overhead trolley conveyor, 22, 219
  sample, 22

P

Packaging goods, as function of
  shipping department, 108

Packing station, 233
Packout equipment, 223–27
  automatic taping, 224
  banding, 227
  box formers, 224
  gluing, 224
  palletizers, 224
  pick and place robots, 224
  stapling, 224
  stretch wrap, 227
Packout time standards, 31
Paint conveyer speed, 32–33
Paint equipment, 216–23
Pallet hand jack, 179
Pallet patterns, 133–34, 164
Palletizers, 224
Pallets, 179
Parking, 142–44
  lots, space requirements for, 142–44
  spaces, determining amount of, 142
Part numbering system, 157, 189
Parts list, sample, 17
Parts, location of, as principle of
  motion economy, 90
Percentage load, definition of, 39
Performance principle, material han-
  dling and, 169–70
Pick and place robots, 224, 228
Picking cars, 233
Picking carts, 135, 228, 229
  tractor-trailer, 229
Picking truck, 228
Pictures, as design component, 13
Pieces per hour, definition of, 42
Pilot, engineering, 310
Planning principle, material handling
  and, 160
Plant flow, 58–67
Plant layout:
  goals of, 3–7
  importance of, 1–2
  introduction to, 1–9
  methods, 284–90
    architectural drawings technique,
      284
    computer-aided design techniques,
      290
    templet and tape methods, 284–85
    three-dimensional models, 285,
      290
  procedure, 24-step, 7
  projects, types and sources, 8
  relationship to material handling, 2
Plant rate, 31–32, 34. *See also R* value
Plastic templets, sample, 288
Plot plan, 281–84
PO operation, definition of, 40
Point of use storage, office design and,
  252
Portable racks, as type of storage unit,
  190
Power and free conveyor, 223
Power roller conveyor, 217
Powered hand trucks, 215
Powered round tables, 207
Pradeo analysis, 128
Presentation, layout, 307–08
Privacy, 248, 251
  office design and, 251

DATE DUE

AUG 9	1994		